The MUSEUM of UNNATURAL HISTORY

The MUSEUM of UNNATURAL HISTORY

Terence Bumbly

Figment
PUBLISHING

UNNATURAL
END FOR UNNATURAL
MUSEUM

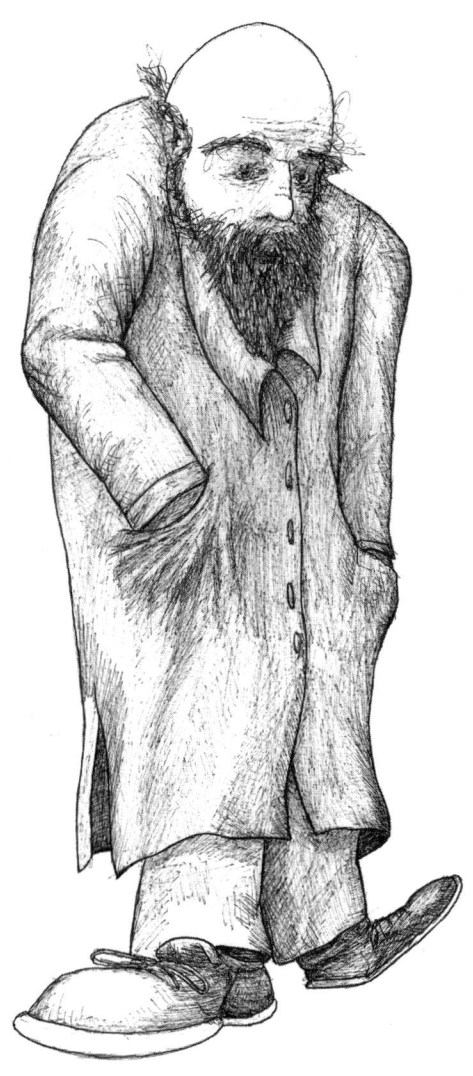

Plate 2:   My estranged brother

Foreword

On my first real holiday since the opening of the Museum, I received the news of its fiery destruction via a local paper; a short column on the fifth page with a small picture and one hundred words covering my work for the last fifteen years.

My estranged brother founded the museum committee shortly before he passed away, leaving me as head of the Bumbly estate, and the Museum of Unnatural History landed in my lap, on a platter.

Not surprisingly an appropriate site with council permissions could only be found in Berlin. So my beloved and I relocated and began collecting the exhibits. The museum was to house the aberrations created alongside the development of the sciences and the arts, intended in no way to criticise our clumsy advancement but simply to note some of the interesting and incredible tangents.

As many of our pieces were still living, we were forced to exist, in part, as a commune. Attracting all kinds of scientific bungles – transient fellows out of place from society, contemporary deformities – we were, for a time, a family of all sorts.

Plate 3:   Terence Bumbly with partner Ms Sveldt

The museum was subdivided into six sections and I follow them here: Breeding, Genetics, Evolution Control, Art Pieces, Mechalogical and Items of Ambiguous Classification (which included the botanical exhibits). Unfortunately the fire left little to salvage, perhaps just a few of the hardier gift shop items.

Since the virtual tour and all the digital artifacts are still accessible on the weave, the destruction of the physical museum will go largely unnoticed. I have chosen to illustrate the exhibits with sketches I had done over the course of the museum's life, as this is intended as a personal account rather than a textbook. Seen another way, I write this solely for myself. Like many elder normals, the need to 'stocktake' one's life events and sum them up into a minor footnote to history is a conceit designed to assuage the fear of indefinite existence, as well as acting as a *gestalt* for what will always feel like unfinished business.

My manner of speaking may be considered coarse for the average logocentric. I have my suspicions anyway about the fallibility of human truth, and more so the fallibility of our language, which may show up throughout the text. I may not use words and terminology as usually understood and I apologize for any discomfort this may afford my readers. On the other hand, I can't help but see all language as a constantly shifting beast, Protean perhaps, that differs from user to user and over time – we can but try to understand one another.

Bumbly

With all due respect
(Terence Bumbly, Former Curator)

CONTENTS

PLATE INDEX

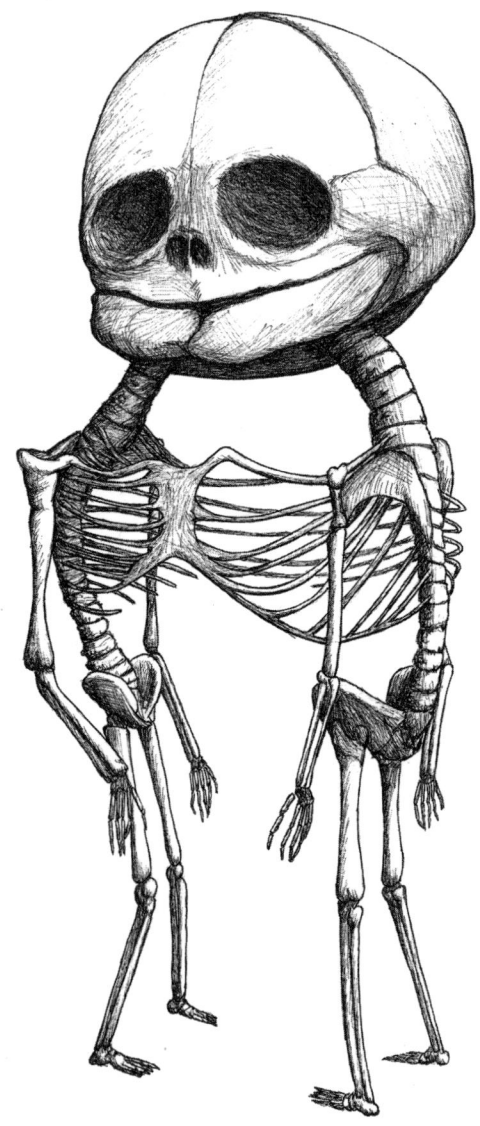

Plate 4:   Siamese skeleton

PREFACE

There are far more grievous offences to the quality and quantity of life on this planet than experimentation. Though such endeavours can occasionally lead to beings that wouldn't otherwise exist, we should view these as blessings rather than abominations. There is, in any event, a long history of natural variation without human interference.

After a long life of witnessing the finest of cultural sophistry, I have an extensive collection of terminology which would best be disposed of. I no longer understand 'human ideals', or 'ideal humans', nor in fact much of human rhetoric. I can imagine no such thing as an ideal form, though I can readily point out some possible improvements in behavior.

Plate 4 shows the skeleton of siamese twins delivered by Dr. Joseph Warrington in the 19th century. Conjoined by a single skull of watermelon likeness and created with no scientific intervention, they and others like them are solely the products of nature, God, or random chance (whichever the reader prefers).

Plate 5: The Feejee Mermaid on the other hand,

is a subject of disputed origin – as are all things we don't fully understand. Real, fake, unreal, imitation? History concludes that it was a fraud concocted between P.T. Barnum and one Levi Lyman in 1842 but people of the time could not be sure for, as we all know, if there is a God, he/she/it has done some wonderful freak-work.

Alas, all natural deformities can now be avoided long before birth with pre-screening, and the off-world birthing tanks have eliminated such chaotic influences as nature to create only perfect specimens of humanity[1]. As such, it was only through the miracles of error and misguided invention that we could gather enough exhibits for our little live-in museum.

1 A contradiction in terms?

Plate 5:   Feejee Mermaid

Plate 6:   Spaceranger Theodor Loom

BREEDING

More historically associated with dogs, cats and horses (i.e. domesticants), there were over the last century many attempts to further the species by this more traditional method.

The pitfalls were the same and it must have been challenging and expensive keeping the gene pool big enough to avoid inbreeding pedigree weaknesses.

As was to be expected, most of the experimentation was done on males in the military – it's practically a tradition now, as we never run out of willing rank and file. Even with cloning as an alternative to naturally occurring humans, the majority created for such tests were inevitably males. If this isn't chivalry, I'm not sure what is.

Space Ranger

The human body wasn't designed for space and hyper-accelerated travel; nor could it survive for very long in the open void.

The goal was to fast-track bodies more suited for zero-gravity that could also withstand up to

20g of acceleration without internal rupturing.

The program used cloning to save waiting for generational maturation and for a while debated whether or not to solve the short lifespan still plaguing the cloning industry. On the one hand, it kept a built-in shelf life as a safety feature against a clone revolution but on the other, it was stopping the program from going after that holy grail, the immortality market.

Plate 6: One of the end results before the funding disappeared was Space Ranger Ted, born Theodore Loom. This miracle of best intentions was a confused, often dizzy, individual.

The ocular advantages of a two-vision field were offset by the inability of the visual cortex to process separate inputs. With vision being so intrinsic to brain development, Theodore was never going to become what they intended. He was, however, excellent with children, painting his tentacular extremities in rainbow colors and performing a whirling dervish act – which no doubt did little for his perpetual vertigo.

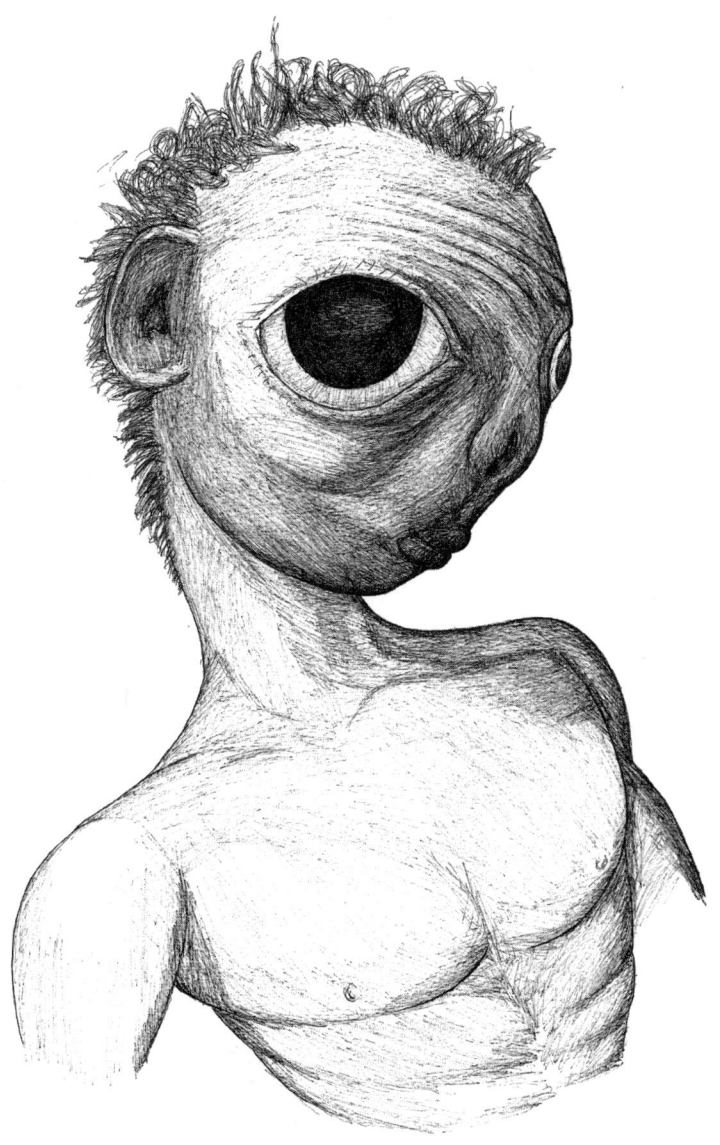

Plate 7:   Offworlder

Offworlders

While not a breeding program *per se*, Offworlders were self-infected with a virus that loosened the strictness of DNA and RNA copying, thus facilitating localized adaptation to the new environment. The results were unpredictable but eventually useful.

Plate 7: An offworlder, of Asteroid Zeta-37 beyond the Kuiper belt. Eyes and sockets enlarged beyond proportion, to soak in the precious and distant sunlight.

Psionic[1] Breeding Program

Until the psionic breeding program, extra-sensory abilities were measured by a random sampling of the population, to see if they actually existed or not. But to find the fastest human in the world you don't test a random collection of people; you hold competitions, events, and make people train. Then, if you offer a big enough prize, almost everyone will give it a try and thus were the origins of the psionic breeding program.

1 For those used to more everyday terms, psionics in this context is related to the practice of a variety of psychic abilities.

Plate 8:   Pierre Jr

The top psi-athletes were then encouraged to breed and spend their lives trying to develop their ability within an enclosed society.

It was, as we all know, hugely successful; which led to their downfall and the inauspicious place they now hold in our society.

Plate 8: Pierre Jr, born of Mary Kastonovitch and Pierre Sandro. An unintentional spawned through extra-marital activities, here shown in his second month.

This tormented baby was the result of two of the most powerful telekinetics ever known. Its brain developed exponentially to its body, to the point where it could no longer move under the weight. There are no surviving witnesses to his disappearance.

I was paid a visit one time, or so I believe, by a very strong controller[2]. She kept me from seeing her, so I can give no report of her appearance. Also, as she never 'spoke' aloud, as such, I can only trust that the experience wasn't entirely imaginary.

Putting that aside, the controller had come in

2 Someone skilled as a central entity for composite minds. When joined for a specific purpose or task, one mind must lead the way, utilizing the strengths of the joined as best they can.

reply to a call I had put out looking for exhibit pieces relating to the psionic progression of the last few centuries. While she brought no such donations, she did bring news about her own encounter with the lost one, Pierre Jr.

He was born a telepath, and to be born a telepath is to be born in the deep end of consciousness, as you share immediately the workings and learnings of all within mental range. For them it is either sink or swim. Most sink, but not Pierre Jr. Perhaps being protected from the hostile and indubitably confusing influences of the larger population and gestating within an enclave of carers, nurses and teachers, performed some sort of positive early modelling on his fresh brain.

Of course all this happened within the womb, swelling his skull, and he was removed via caesarean. His psycho-kinetic potential was beyond control, as too his mental domination. As such, none could stop him escaping when he chose. He was the boy who could levitate before he could crawl.

The feeling of having your mind probed is like suddenly standing after having your head between

Plate 9:   Vitreous headgear & Ganzfeld Hood

your knees, or as if someone has plucked your mind like a dandelion and is blowing off the seeds: a most disturbing helplessness that is over at once, and easily dismissed as a passing blood rush. We all have such moments and as she let go of me I was confoundedly unsure of my memory, although I suddenly had a much greater knowledge of the psionic world.

Once it had been proven that telepathy existed and that *some* people could enter another's mind, the issue of security came to the fore and many dubious contraptions were devised.

Plate 9 shows an example of the kind of productions created to counter anxieties over the psi threat, or more accurately the *possibility* of a psi threat. All the most important people wore such headgear; politicians, judges, pilots and so forth, people who needed to have a 'clear head.' It was claimed that the vitreous helmet shown here would block any attempted breach of the sanctity of one's mind. How naive it seems our forefathers were. I don't believe any such contraptions had any effect on psi-ability.

A similar device could also be obtained to help

enhance psychic potential. Notably, the Ganzfeld Hood which blocked external interference and sensory information so the wearer could detect, and thus exercise, their latent abilities. Again, I'm not sure it worked to any extent.

Psis now work the black market, giving clientele experiences and memories that cannot be achieved elsewhere or elsehow. It is highly illegal and therefore highly lucrative – but since any practitioners must be able to read minds it is unlikely they would reveal themselves to any who would turn them in. Some may be altering our perceptions so they could never be identified anyway. Who knows? Depending on what degree of subjugation they can inflict they could be controlling the law-makers. Though it could well be futile, the laws will remain in place until we adjust to the new possibilities.

The fear now goes that telepaths reside in recessive personalities and are thus undetectable until they surface, and to increase their numbers, and numbers of sympathizers, they could be posing as maternity nurses and school teachers and so manipulating young minds.

But it is not a war: it is an evolution, and for

those of us who missed that evolutionary leap, there are now communicator implants that use neural interception and intervention techniques. It's not the same blending of minds that comes with true telepathy, but a reasonable substitute.

The second big fear is regarding direct human communication on a mass level. Consider firstly that everything has ranges; of ability, quality, quantity ... strength? I don't know the words for such things, but imagine that each psi has a certain strength. This therefore leads us to consider a dominator archetype, who can remotely overpower the thoughts and thinking of the weak-willed.

The good thing about traditional communication is that you can switch it off if you don't want to listen – this may not be the case with psis. If such a dominator came into being – a malevolent personality such as Örj[3] for example – where would it end? Could we become like bees, with a centralized 'hive' consciousness?

I shudder at the thought, though our heirs may feel differently. Those new to a changing world always have an advantage, and so experience is outplayed by ignorance.

3 See page 19 for those who don't know their history.

Plate 10: A canary with a sore tummy

Canaries

Like their extinct avian namesakes, 'canaries' were bred as an early warning system for epidemics. Criss-crossed with canines, felines and all sorts of mammalians, including Homo sapiens[4], came the breed of animal we now call canaries. Their weak immune systems and accentuated tell-tales (such as bright red noses and sweaty-ears when sick) could provide enough warning to medicate and protect their human owners from suffering the same condition.

Canaries are the most popular domesticant on all the worlds and throughout the fringe, safe-guarding the health of more privileged creations. We can be such monsters.

Breeding is considered a 'normal' process, attuned with Darwinism, though it is controlled and directed through human selection. Perhaps this means we have been beyond Darwinism for millennia.

---

4 Incidentally, "modern" man is a further sub-species called Homo sapiens sapiens, though I don't think this matters here.

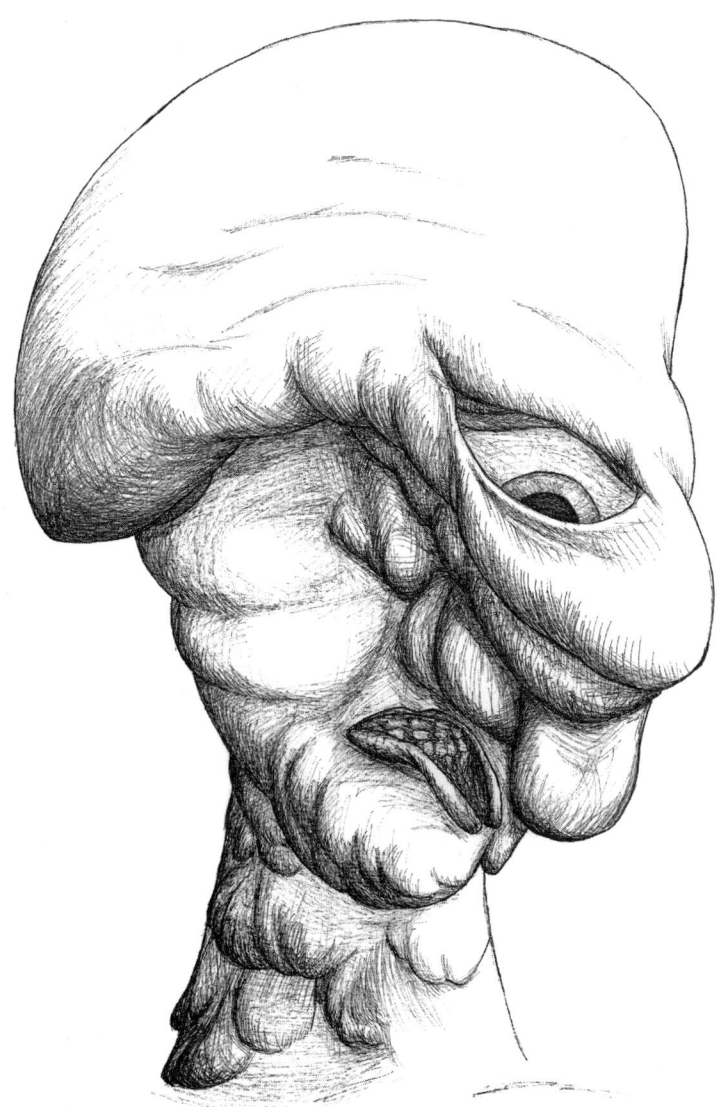

Plate 11: Preposterous growth

(Eu-) Genetics

There were simple aims to genetic modification and enhancement but there always seemed to be a directly proportional relationship between the intended benefits and undesirable side-effects.

Genetic bending was largely illegal and thus the greatest advances were made by countries external to the World Union (WU), which only placed the limit of 'doing something useful' on scientific investigation.

So in an effort to do something productive for these second class countries, many ingenious traits were accentuated and amplified. Unfortunately, due to the complexity of natural interaction, the side-effects were unpredictable: bizarre enlargement or foreshortening of limbs, conjoined creatures and other distinct features.

The most curious results of genetic experimentation were on non-human fauna and flora, which we shall look at in the garden.

Plate 11: When cataclysmic pollution levels swamped the planet and its populace with health problems, scientists desperately tried to create a human with a much higher chemical and radiation tolerance. But what is accelerated in the tomato becomes preposterous in the human.

While this direction became redundant on Earth after SIB[1] control was implemented, this breed has done well on planets like Mars where a thinner atmosphere lets in more cosmic radiation.

Eugenics, Eugenetics and You

Words, words, words. In my lifetime I have finally swallowed the hollowness of words, or perhaps I mean the weakness of them.

Debate never ceases to rage over what is right and wrong, good and evil, etc. – but manipulating what is deemed legal or criminal serves in no way to change the behavior of people, only the classification of such.

Laws and condemnation do little to sway the human animal. All attempts to curb destructive inclinations and acts deemed negative are, by and large, flawed unless they confront the basic factors of influence. Unfortunately I can only extrapolate

1 Superior Intelligence Being

from personal observation and say that we are guided by example, etiquette and tradition, though emotion and need can effectively bypass our normal *modus operandi*.

The debate over manipulation, especially in the areas of breeding and genetic modification (since they have the highest 'failure' rate), has only served to push these practices underground, making conditions for the experimented even worse, not to mention that off-world policing has proven quite impossible – if you don't like the rules on Earth, there are only financials stopping you going elsewhere.

My earlier belittling of words aside, they can become abnormally powerful if they get dirty. And here I come to talk about eugenics, a word so dirty it was practically wiped from the records (which could happen more often than we think since there would be no proof of a successful wiping).

Eugenics itself was practiced world-wide before the Nazis[2] made it a dirty word, and as such all references were changed; textbooks rewritten,

---

2 Nazis, or Nazism succeeded the failing Weimar Republic. Largely defined by the ideology of Adolf Hitler who led Germany into the second World War, and slaughtered people on the pretense of superiority.

scientific journals renamed, and strong denials echoed all around.

With our incorruptible belief in miracles, humanity convinces itself that uncomfortable concepts can be bypassed simply by changing the words used to describe them. I'm not convinced: I believe in calling a spade a spade, if not a dirty old shovel.

At its origins, eugenics was the study of the self-determination of human evolution. By whatever means, with whatever parameters or aims, the 20th century attempts were both monstrous and clumsy. Still hung up on 'ideal' concepts of humanity, governments tortured their populations with voluntary and involuntary sterilization, mating restrictions, segregation and extermination. There were some noble aims such as eliminating certain hereditary diseases, but the strange dogma of the time considered that race-mixing led to 'impurity,' and that civilization itself, by looking after its weakest members, was going against the evolutionary goal of only the fittest surviving (which wasn't how Darwin originally phrased it).

Eu-genetics, the modern counterpart, quickly branched off from its ancestor, not just in scientific method but in its aims. Eu-genetics does not divide its categories into positive and negative (which are socially and culturally determined), nor does it aim for homogeneity. Rather it encourages diversity as the real strength of the species.

As a privatized industry, it also lacks the coercion of national schemes and is motivated by individual competition and the desire to give one's progeny all the advantages money can buy. Here, of course, it reinforces the economic divide as the rich can afford to make their children far 'superior.'

The current rhetoric of 'conscious evolution' – as well as claims that it is merely technically assisted evolution – seem eerily similar to the sales points attached to the original eugenics, but to them I ask: What's the rush? Where do we think we are going anyway?

Örj

Every museum and history has a section dedicated to Örj, and we were no exception. Örj the Consumer, Örj the Great Descendant, Örj the Inheritor and Örj the eater of his own progeny.

Beginning as an academic, Örj was a great

Plate 12: Örj incarcerated

believer in the path of evolution, with a particular fascination for some Lamarckist[3] notions. His self-experimentation sullied the ideal of technological enhancement for over a century as his attempts at self-improvement went far past stimulated changes – he also aimed to make those improvements hereditary.

Knowledge is power and power corrupts, as they say. His followers were many, sweeping down from Norway in hideous forms to promote, if not enforce, his new gospel.

Örj sired a great many children, with and without the consent of the host mothers (many of whom suffered fatally from the abominations gnawing inside their wombs).

Plate 13 shows how successful he was in creating children in his own image. Kyla Örj was an excellent chef who invigorated Scandinavian cuisine by perfecting the re-creation and cooking of the native reindeer.

Though only a small percentage of his offspring survived gestation, and even less the early years of childhood, a good five thousand reached adulthood

3 Jean-Baptiste Pierre Antoine de Monet, Chevalier de Lamarck (1744 – 1829), French naturalist

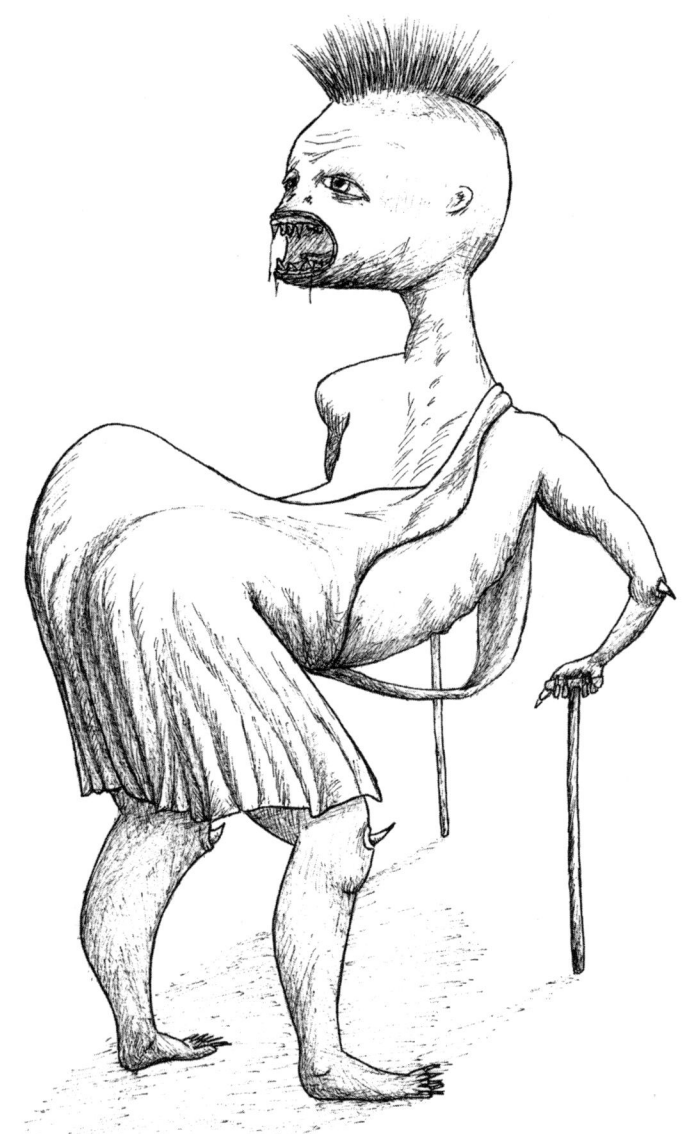

Plate 13: Kyla Örj, daughter and grand-daughter of Örj

whereupon the Örjian tradition of combative evolution pitted them against their own father and any other who sought to climb the food chain.

Luckily, few survived[4]. The Örj rule of thumb was that the son must better the father, or die trying, thus increasing the level of competition and strengthening the gene-pool, supposedly leading to the betterment of mankind. Humbug.

I, of course, dispute such 'thinking.' By focusing on competitive evolution one simply makes more emulous and violent creations, and discounts the evolution of civilization as a whole – which is where the real progress has always been.

Perhaps I should have slotted Örj and his ideas into the following exhibit on attempted evolution control, but he always exemplified the motivations of eugenics despite the artificial enhancements. His methods were similar.

Distinctions and classifications are arbitrary creations, forever changing and shifting with the whim of the observer. So in the spirit of splitting

4 There hasn't been such a monopolization of the gene-pool since Genghis Khan, so it was 'good' that so few remained to spread his opinions of humanity.

hairs, 'attempted evolution control' is more about changing the nature of evolution itself, rather than its results.

ATTEMPTED EVOLUTION CONTROL

Otherwise known as forced-evolution, this section separates itself from breeding programs and genetic manipulation mostly in the motivations for what was created. Genetic experimentation was aimed at perfecting what already existed, while breeding was intended to fast-track directions in development. The various exhibits represented the intentional redesigning of humanity into new forms, to in fact step beyond the bounds of 'human.'

Of course, the questions of 'what is human?' and 'what is not human?' again reared their dull visages for the millionth time in history. I have always found such questions miserable, linked to arbitrary categorizations such as DNA profiles or behavior – not to mention tedious theological poppycock.

What do we expect by asking such questions? Since everyone has different answers, we get arguments, dichotomies, and hierarchy. Are some people really asking, 'Who is better than whom?' I think so. Perhaps if we started with what helpful effect we wanted the answer to ignite, we could just work backwards from there.

With more frequent excursions into space, the limitations of the human body were quickly encountered. Evolution creates tight bonds between an animal and its habitat, and it seemed easier to change the human creature than to recreate the necessary environment in space.

The term 'adapted species' intruded on our vernacular: humanity, re-sequenced, bred and hormonally enhanced. Those damn Prometheists[1] like to use the terms *rudis* (raw), *correctus* (corrected) and *altus* (improved). Hardly anyone outside of the movement uses the terms, unless to distinguish between the Promethean ranks.

There are some who take a wider view of evolution than just the changes of the physical body. Few modern humans would consider themselves in regard to their bipedal posture, recessed canines and enormous brains (respective to body mass). So perhaps it is integral when considering evolution to think of the knowledge, skills and social structures – perhaps anything 'of humanity' that has ever

1 See page 51.

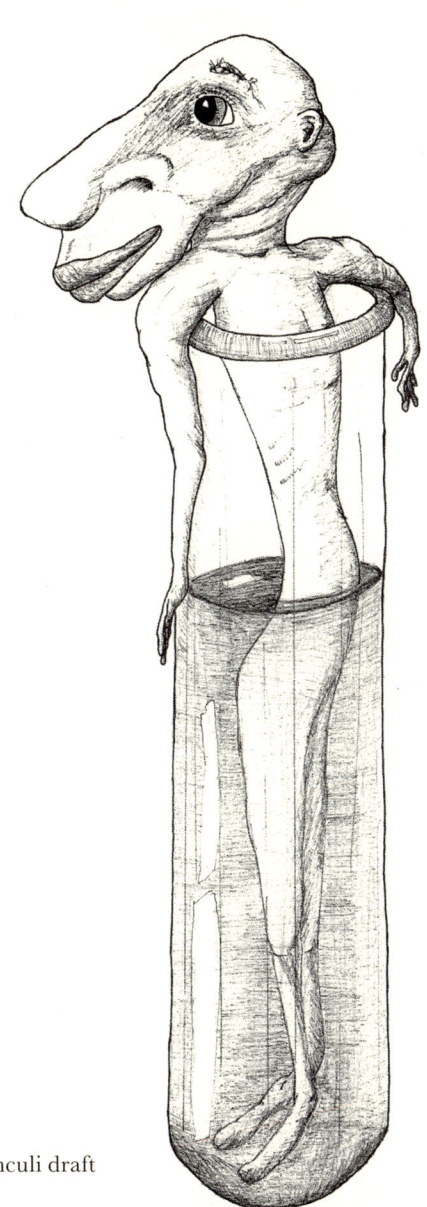

Plate 14: Homunculi draft

changed or developed in any way.

Here we begin to place symbiotic technological relationships into the evolutionary sphere. There are already such products that grow up alongside human children, and even inside them if we start talking nano-tech.

As with any development which shuffles human equality, the ability to alter and improve on your *au naturale* human has created a separation: the useful and the non-useful. The non-useful, such as myself, lead our humble lives idling away at whatever distractions are available to us. The useful are more often than not specialized creatures, so tied to their function that they hardly socialize or breed.

The non-useful remain as a safety-net in case all the biological meddling leads to a dead end; also as a pool to draw from to create more usefuls if needs be. I still don't know who decides what needs be, what needs need be, or what needs be needs, but maybe it is obvious to the higher-ups.

Homunculi

The first miniature humans, or homunculi – their

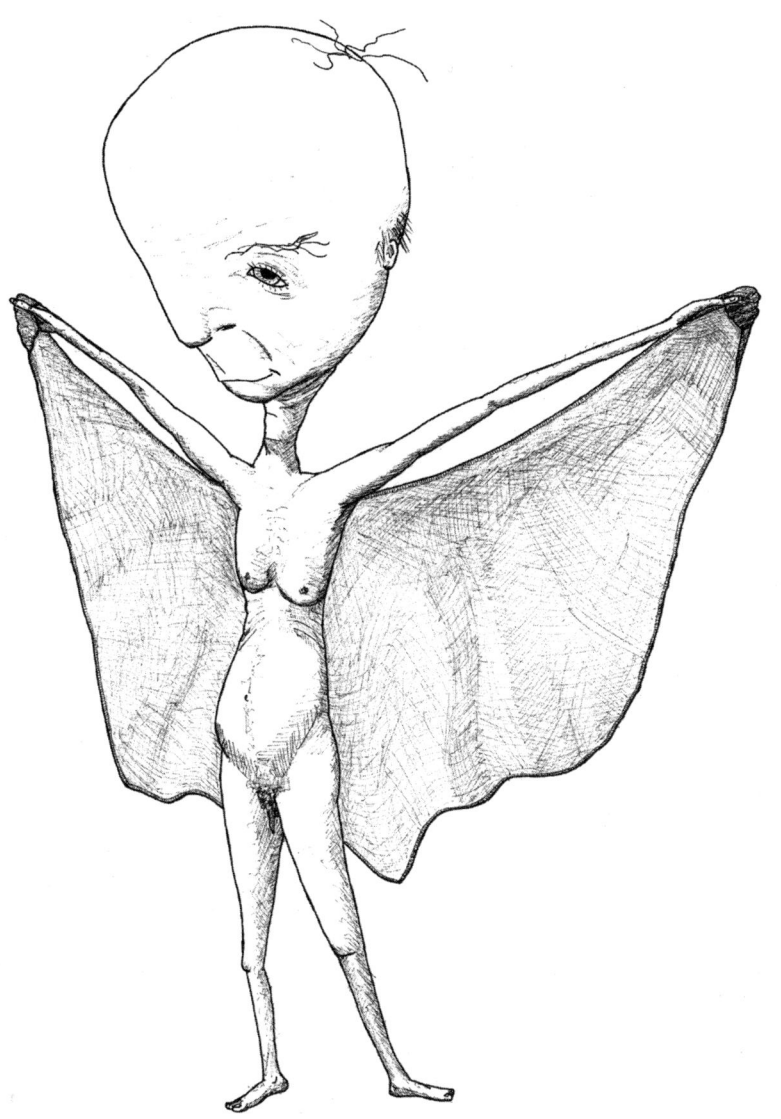

Plate 15: 'Behold, I am the greatest among you!'

bones too weak to support them (Plate 14) – were nonetheless a vital step in solving the problems of over-population then suffered on Earth. On average, humans are now one third the size they were at the beginning of the 21st century.

Delusions of grandeur frequently affected the big-brained little people, though often with more than a hint of self-mockery. 'Behold, I am the greatest among you,' one peculiar chap would introduce himself to any group, gloriously pulling open his blanket (Plate 15). This one lived most of his life in quite a small asylum.

The combination of smaller bodies, advanced intelligence and flight was thought to be the next step in controlled evolution. The result, though, was a rather anti-social sub-species with little interest in contributing to human affairs. Some suggest they feel ashamed to have been created by humans. So the question for homunculi must be: what is it to not be human? And yet not alien or robot? To be of human, though far superior in all ways…?

We'll never really know since they quickly created a new, presumably improved, language – unintelligible to us, for their new, improved

Plate 16: Homunculus escaped!

thoughts. Assuming of course that 'thinking' has not been superseded by whatever it is that they do with themselves.

The few homunculi that gathered together in the museum stayed merely for protection from their creators. I once asked them what they had to fear from their creators and one deigned to respond in my own language, 'Our creators were dedicated scientists; it is those that seek to use or harm us we stay away from.' A finely made distinction. (Plate 16).

Mangutan

To avoid ending this section on a dour note, there were, at least in my opinion, some successes from the experiments in controlled evolution. It was ironic, or perhaps I mean poetic, that it was a mistake that created them, thus uncontrolled evolution. The mangutan, or humangutan to reduce gender-bias, is thought by some to be a throwback to pre-homo sapien status although it is quite distinctly on a separate genetic path to modern humans.

It was via experimentation into human origins, and investigation of our ancestral forms that the

Plate 17: Mangutan

mixed-up breed of the humangutan originated[2].

Retreating to a standpoint of cultural-evolution through spiritual and personal improvements, they have shown themselves to be the most lovely of creatures and perhaps set a new example for humankind.

They are an almost languageless creature, at least so it seems. Scans show enormous amounts of brain activity but they restrict themselves to well-toned and impeccably timed grunts. They could speak words easily enough, though seem annoyed by the need to do so − more a failing of us than them. I suspect they rely more on body language and empathy, if not to communicate, to predict the meanings of other mangutans if they were to speak out loud. Perhaps it is best to consider it a shorthand speech.

I have since travelled to the mangutan forest, where they drifted together. During my time there, I could swear I heard the sound of homunculi wings and the tender scent of screaming head orchids. I wonder how many of my old friends escaped from the fire and now hide where they can.

2 Using remnants of *Australopithecus* and *Neanderthal*, with sequence gaps filled with modern human cells.

Some ask if all these technologies and changes are endangering the human species. It seems to me that only the word 'human' is in any danger. Within a few generations it may only denote a historical form or 'birth form.'

The only people who would be offended by such a loss would be long dead, their opinions only for scholars, much like medieval belief systems are to us now.

On the other hand, people who worry things will change in a day, or a lifetime, haven't looked over history to see how slowly things change. I don't think human ideals and concepts will be thrown aside so easily.

Plate 18: The Pianist

ART PIECES

One of the infernal questions: what is art?

Well, 'art' is a word that was once upon a time applied to such things as paintings, ceramics and sculpture. Subsequent generations asked, 'No, really, what is art?' and the word garnered near mystical speculation over its essence and true meaning. After centuries of that we are more likely to ask, 'What isn't art?'

Plate 18 will offer no illumination. The pianist, Strungoff Vortex, whose asteroid-based hermitage suffered a system failure and slowly radiated over an 8 month period until what happens to all of us eventually happened to him. On one occasion a child erected their cracked lollipop into the head of our pianist. The parents were suitably mortified but we all found it rather funny, so we left it in.

Accompanying the exhibit we had the music Strungoff recorded while he died (and a few preceding months for reference's sake), which is a lifetime of listening. His highlights compilation, 'Music Recorded While Dying of Radiation', was a popular novelty present for music aficionados.

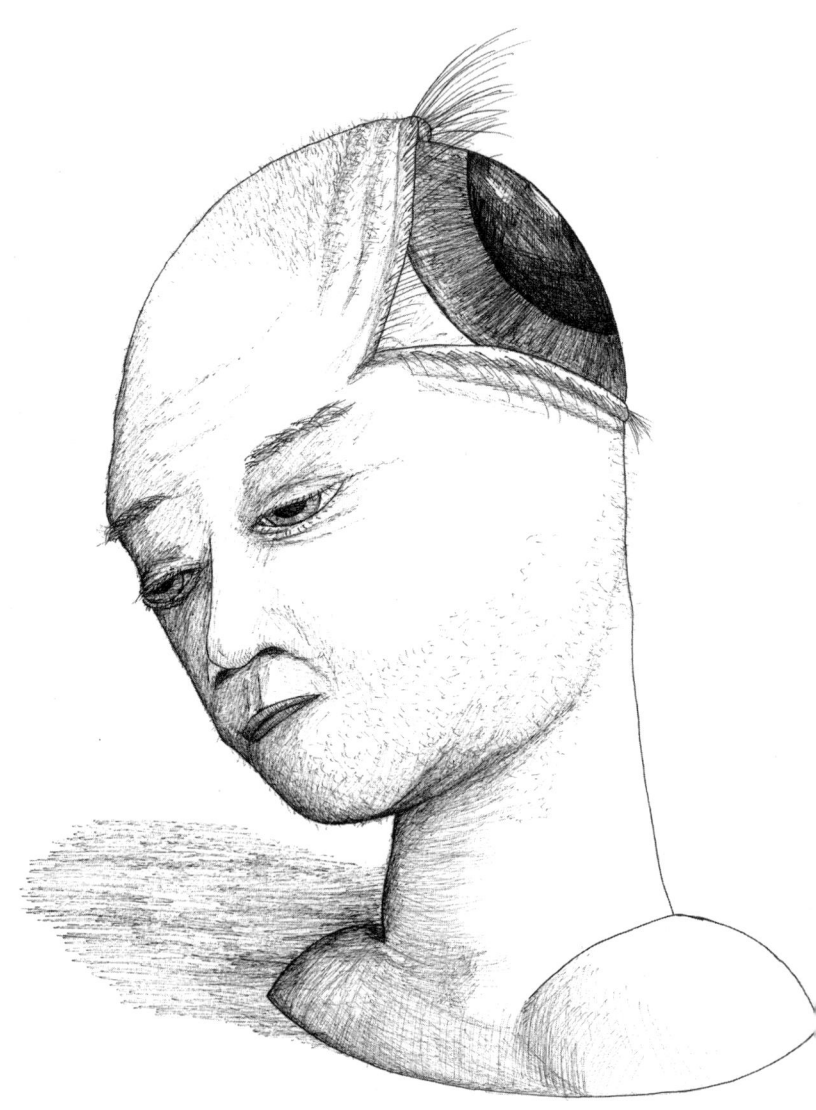

Plate 19: Lorret's new vision

Art Pieces

Whether considered 'good' or 'bad', art reflects the artist, as well as the reality and zeitgeist that surrounds them. There are times when art makers desire intrinsic meaning, or prize virtuosity, or need money and see a way of getting it. Value is often found by those who identify most closely with the artist. Thus, alas, art returns to the eye of the beholder.

What seems ridiculous about these meandering art arguments is the initial idea that there could ever be universal agreement[1] when even individual words can be hotly contested and skewed by personal context – how could visual stimulus be bound to some conformity of understanding? That would require a conformity in thinking and perspective; the removal of the individual in fact.

One could apply that theory to any discrepancy between people – but that doesn't mean we shouldn't try to agree. And thus art and discussion have persisted; perhaps the human essence lies between rather than inside us.

1 Or even just on Earth!

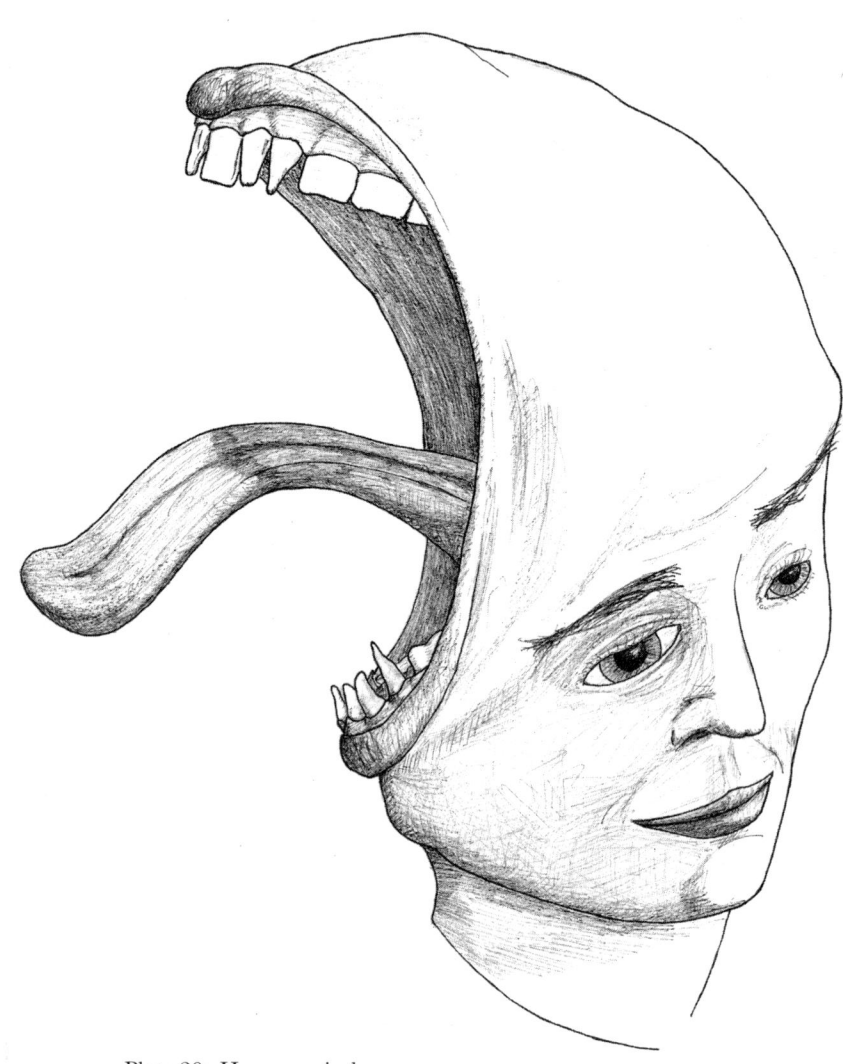

Plate 20: Hungry mind

Ernst Lorret

Some suggest there is nothing new under the sun and all art is just a remake of previous ideas, that there are only a finite number of basic concepts and that we are forever destined to plagiarize. Such calumny, I dare say, indicates more about the speaker than what is spoken.

Ernst Lorret (2134–2201) was inclined to look at the matter in reverse: that in fact there are only new things, no two the same. Even a copy is different from the original (in time, personal perception and other subtle differences). This from Lorret's diary:

> Of course there are common elements in art! Everything in the universe is made of the same matter, but you don't go around saying it's all the same [profanity removed by editor] thing, do you!?

Speaking only for myself, Lorret's pieces were initially disturbing, but over time I did come to associate them with a certain freedom of mind, as they illustrate the notion that everything is possible.

They strive for hideousness, yes, and the grotesque, surely, but was this not Lorret's reaction

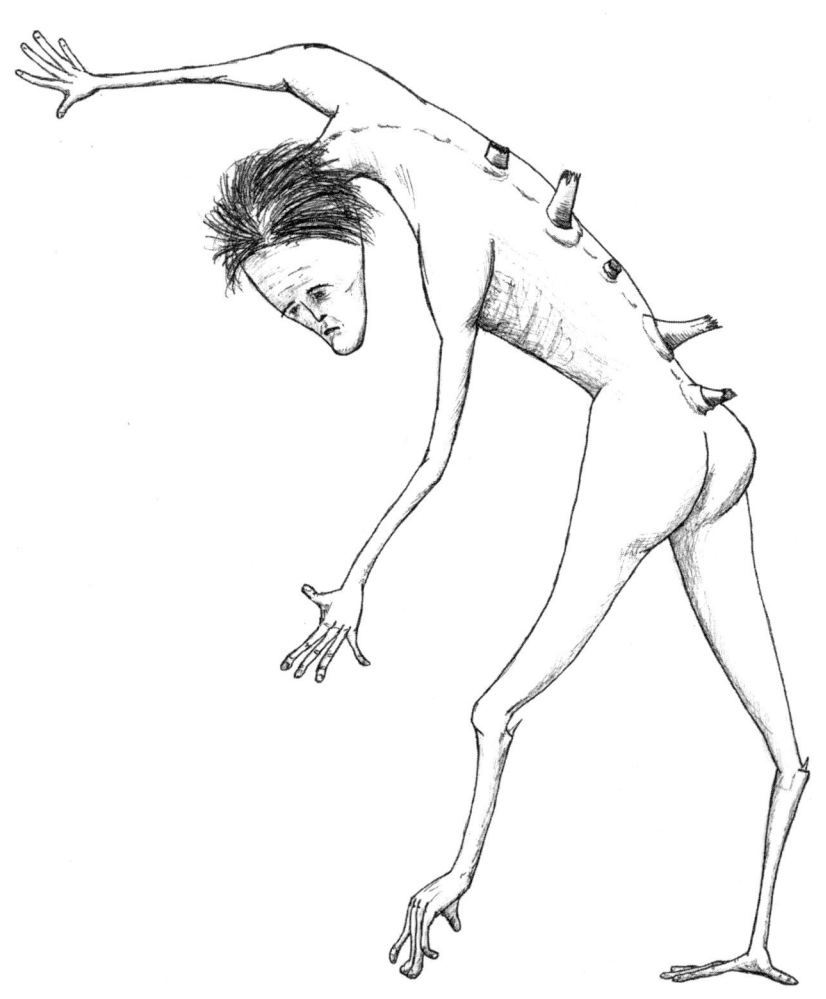

Plate 21: Escapee

to the beautification of his parents' time, and maybe a reflection of contemporary self-disgust and apathy?

Lorret had two methods of working. Besides random experimentation, he would also draw hideous sketches of possible sculptures and then try to create them by any means necessary. Lorret's second approach was to grow 'normal' pieces and dement them by various methods. His abilities to keep them 'alive' after his creative surgeries, showed him to be a genius physician and some of his techniques are still used in hospitals to this day.

He was the pinnacle and driving force of the deformist movement, sometimes referred to as 'Our Deformed Father' in satires.

Plate 19: One of his first and most successful pieces. I will remember it for the weekly shave it required while still alive – a most disconcerting experience.

While living, the pieces needed constant monitoring, especially the example shown on plate 20, which would throw itself off balance if the tongue from its over-sized maw swung around too much. Incidentally, Lorret made sure never to give

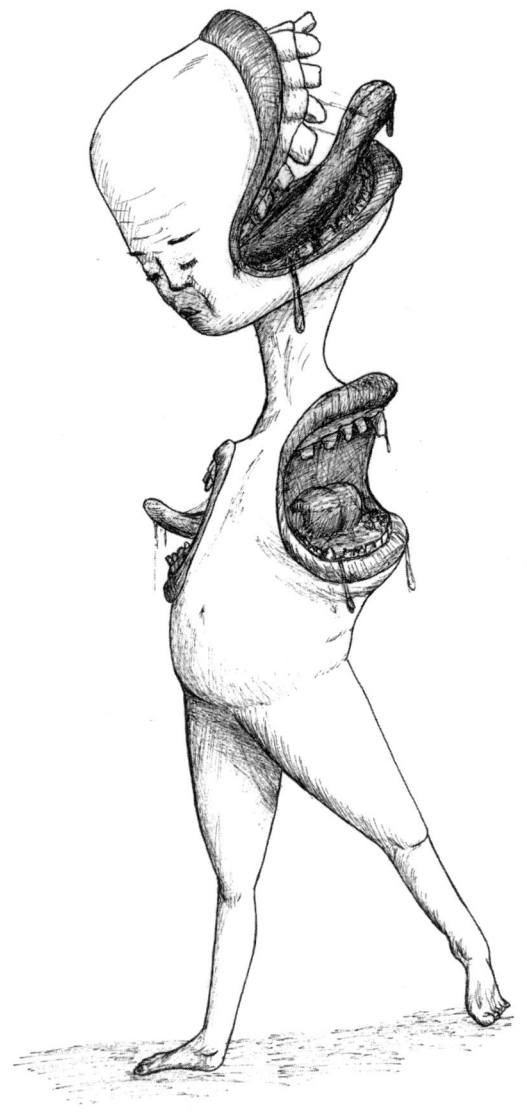

Plate 22: Becoming self-derivative

his pieces vocal cords, so they were silent except for the dull thuds of their toppling over.

Plate 21: An escapee, broken loose from some hideous living-art installation. When art began exploring new forms of expression, tortured life-forms were created practically willy-nilly, without much thought for the humanity of humanity. But then these were troubled times and the new sciences and over-population had displaced the value of singular organisms.

Plate 22: After a time all artists either become self-derivative or begin to explore, usually unsuccessfully, new expressions. Lorret, by the end of his career in the limelight, must have felt he'd explored all the combinations of anatomical rearrangement he could achieve. He could shock and surprise no more, and gracefully bowed out, becoming a patron and supporter of the electro-noise trauma-trance experiments[2].

Plate 23: If art is a form of self-expression, then we can divide them in a number of ways. There are artists who intimate that there is an essential meaning and purpose to human existence, and others who demonstrate that there is no point to

2 Where catalepsy represents the highest form of human experience.

Plate 23: Altered while living method

anything (especially their art). In-between we find a group who make it a ritual to construct a meaning and design to our existence. I like to think that Lorret was poking holes in their fabrication.

While at the time the Museum existed, most of Lorret's pieces weren't technically 'alive,' we displayed footage of the times when they were. With all their blinking, yawning and drooling, their life span was mostly under a decade long – artificially maintained since few had digestive abilities. The recreations we had in the gift shop were popular items, especially the pencil sharpeners.

Madumbe

Found in the outer reaches was a strange device unlike any known human invention. Upon the application of any slight heat, such as the fingertips, it would flicker and fuster into life and project a scant few minutes of footage. The images showed what seemed to be an alien race with a strange crackly soundtrack of slurping noises (Plate 24).

Reminiscent of the Voyager spacecraft of the 20th century, this similarity of idea could suggest

Plate 24: Madumbe

that the recording and accompanying device are merely elaborate fakes. This question of origin left it placed with the art pieces, since either way it is an artifact, whether alien or not.

There is a hypothetical window of opportunity for contacting an alien civilization[3]. In essence a very short window of mutual beneficence. At one point they may be too young, and a visitation from the stars could be confused and construed into a mythological event. On the other hand a civilization too far advanced may be beyond our comprehension and more than likely we would be reluctant to approach them as inferiors. Alas, perhaps we ourselves are beyond being interested in such things, as the source of the Madumbe recordings has never been investigated.

With the time pressures of space travel, the likelihood of reaching a sufficiently advanced culture, after discovery of their existence, but before they develop beyond our understanding is slim. It could take 300 years to send the question and another 300 to receive an answer. What question of any value could you really ask?

---

3 Since non-Terran lifeforms are only theoretical, any supposed science surrounding them must be super-theoretical.

Plate 25: Those damn Prometheists!

Those Damn Prometheists! [4]

The world's first self-acknowledging religion. That they accept their myths as fairy-tale and fiction, does nothing to stymie the fervor with which they preach their beliefs. For them, a supernatural creator has been replaced by a supernatural future.

The Prometheists have established the largest virtual arena on the weave[5] to 'test' hypothetical human models. This 'church arena' is accessible world-wide for believers and browsers to plug-in and – if you'll excuse me – 'fantasize'.

A common symbol of the Prometheists was the third eye[6]. As a gift shop item, the kids loved them (Plate 26). They blended seamlessly with the skin and, powered by the wearer's body heat, could blink, follow motion and dilate with changes of light.

The movement is currently under investigation by the consumer agencies that govern religious

4 Prometheism – a transhuman religion that has done away with 'creator' and 'deity' concepts. Identifying with the ancient Greek icon of Prometheus, they preach the transcendence of the human through technology. 'Transcend to what?' you may ask; Post-humanity of course.

5 Earth-based data network.

6 Appropriated from the archaic symbol of enlightenment.

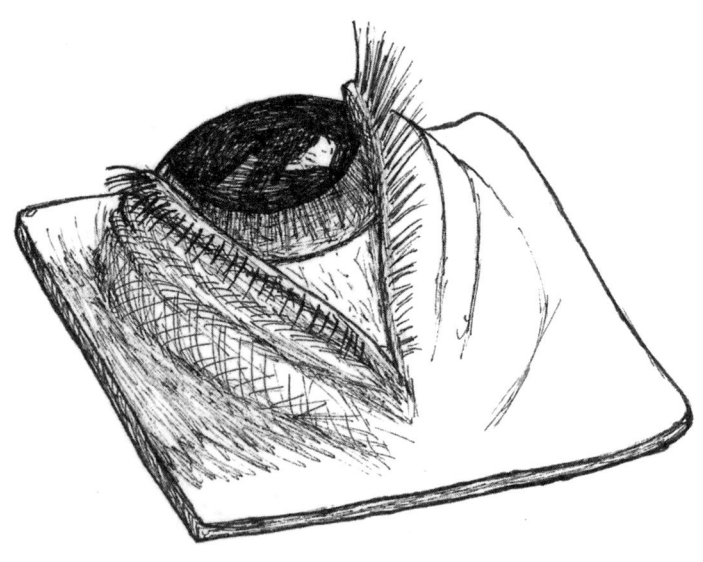

Plate 26: Third eye-patch

groups, the entertainment industry and other 'belief manufacturers'. By taking as gospel the idle speculations of science, there are doubts over the sanctity of the 'physical laws' that they advertise as operating within their online church-arena (thus participant's fantasies have no hope of being attained outside the virtual existence). As such they may be re-classified as an F provider[7].

The other dilemma now facing the transhumanists is that regarding the early practitioners who, three hundred years ago, put themselves into cryonic suspension, to be woken when their post-Darwinian world had come about. This is the longest time any creature has spent in sustained suspension and there are fears that the longer they remain frozen, the less likely it is they could ever be re-animated. Is this the world they have waited for? Will it ever be? Should they just dispose of them quietly? Since there are only a thousand of these human icebergs, the loss would be but a drop in the ocean of the current population, and as evils go it would hardly compare to past atrocities.

7 F for fantasy. All creators of artificial sensory stimuli are subject to Realism Classification, their RC rating then determines how they may advertise their products. It would be a severe blow to the Prometheists to lose their P status (Possible).

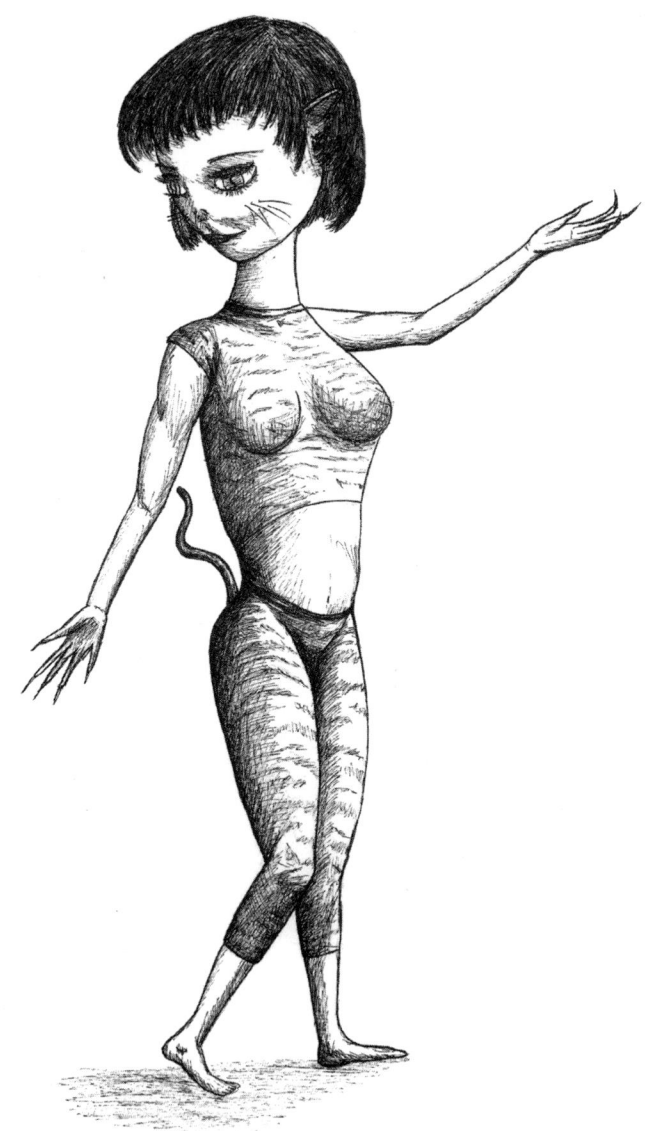

Plate 27: Ms Franjipani Cat

Frea-Kreations™

It is truly amazing the lengths people will go to personalize their bodies. Frea-Kreations™ and copy-cat companies could grow or surgically attach all sorts of appendages and animalia. The Japanese company's bizarre augmentation studios created an avenue, for those so wanting, to be 'fantasticized,' as I believe it translates.

Far beyond the transfiguration and dysmorphia of centuries past, a significant portion of an entire generation became fixated on becoming living avatars, at least symbolically. Freaking has remained the most popular form of 'self-art' for over a century.

Plate 27: Ms. Franjipani Cat, one of our lovelier tour guides, and reason for much passing trade. Formerly the poster girl for Frea-Kreations™ and star of many sensuals[8]. As expected she aged like plastic, with scratches rather than wrinkles. Nevertheless she was still rather lovely and after the museum was destroyed made her way to the Fringe, where she remains popular.

Plate 28: Heshe, the girlboy who was illegally

8 Like cinema but visual, audial, tastible, smellable and tactile – the genre of such sensuals is akin to the literal meaning of the word.

Plate 28: Heshe

freaked by his parents[9]. Heshe was of course too early in physical development and outgrew some of the living augments to the detriment of the implants and the host. There are tougher restrictions now, but you can always go off-world for that sort of thing.

I would show more examples, but such body alteration is common-place and can be found out and about on the streets today as the change-culture persists in more and more glorious forms.

Masquerade and Erotics

In line with the push for human re-design, came a flood of products to support the period of Libertarianism we passed through in the mid to late 22nd century. The erotics industry rose in prominence and legitimacy, catering for all tastes without question.

Popular in the infancy of the culture were living masks and full-body suits for masquerade gatherings and fetish use. Plate 29 shows one of the earliest living masks, which was purely functional

---

9 We are legally obliged to make clear that the practitioners who oper-ated upon Heshe were in no-way affiliated with Frea-Kreations™.

in mid-21st century cities to cope with pollution and the need to breathe. The habitual wearing of such may have pushed the masquerade culture as citizens grew weary of looking so drab because of mere smog.

We also held a small collection of first-edition and prototype passive stimulators, shown here on plates 30 and 31.

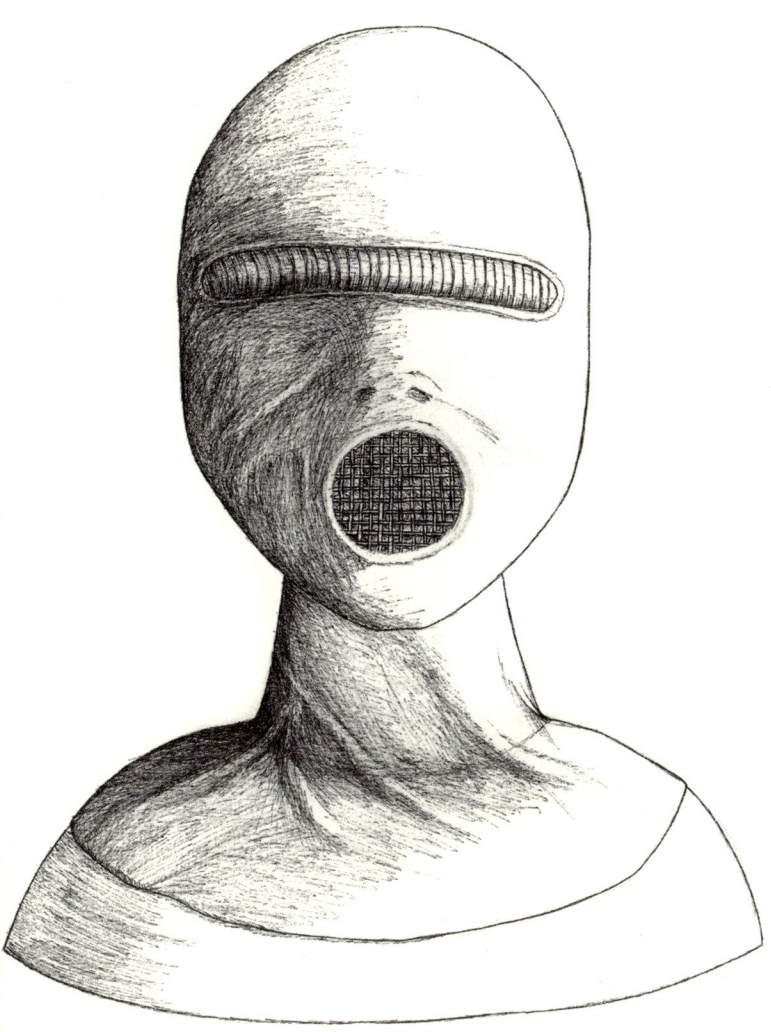

Plate 29: Living mask

Plate 30: Passive-stimulator

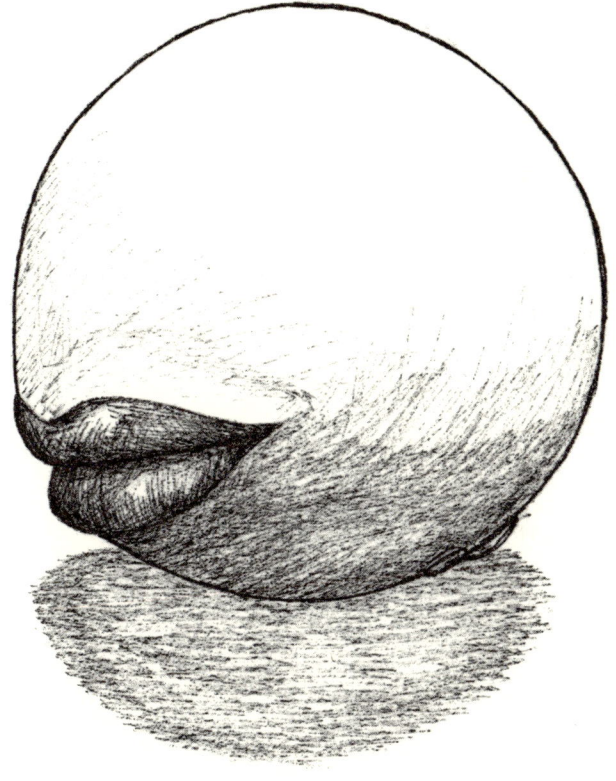

Plate 31: Hand-held passive-stimulator

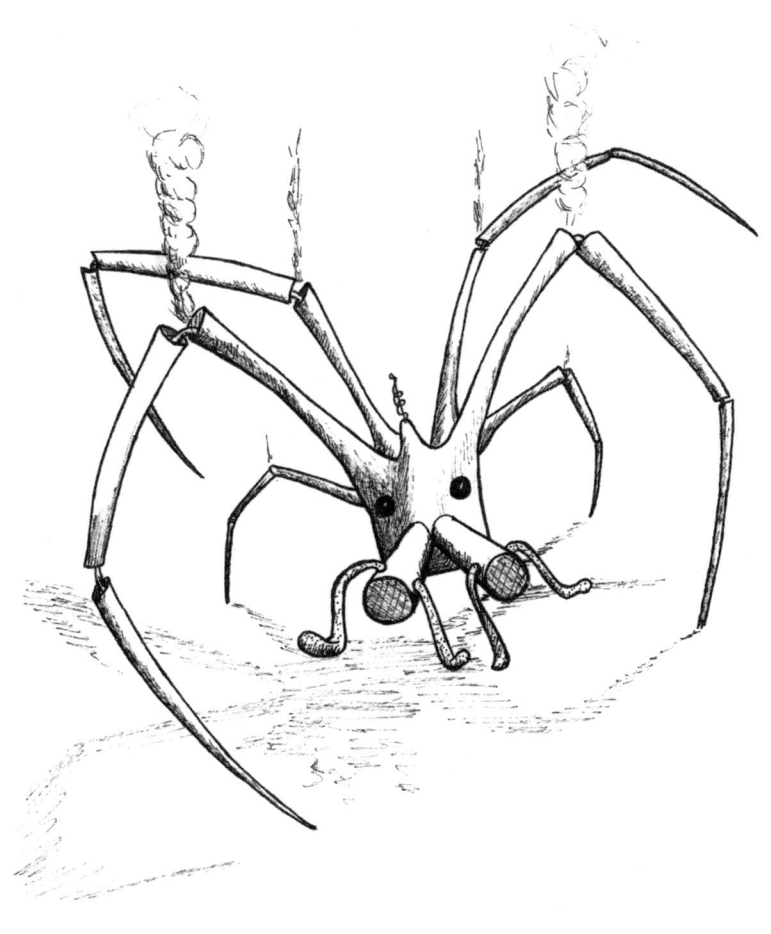

Plate 32: Extinction machine

MECHALOGICAL

The term 'mechalogical' is probably my own creation, but I didn't want to create arbitrary distinctions and subsets between the poles of mechanics and biology. Whether it covers an optical replacement, cybernetic extension, nano-inception or the hypothetical SIB IV – this classification merely indicates that here the twain have met.

Plate 32: an old extinction machine. You could enter any number of presets and filters to specify the characteristics of the animal you wanted rid of, including humans. With its array of sensing proboscises it could track down all creatures of a particular genetic strain and even details such as if they parted their hair to the left or right: a truly amazing example of human idiocy, cruelty and ingenuity in one product.

Computer intelligence didn't come about as many predicted, that is, via a reverse engineering of the human brain[1], but rather from radical storage and processing techniques such as IntelloPlasma™ and InfoRec™. Consequently, there wasn't a

---

1 If imitation is the sincerest form of flattery, what is self-imitation?

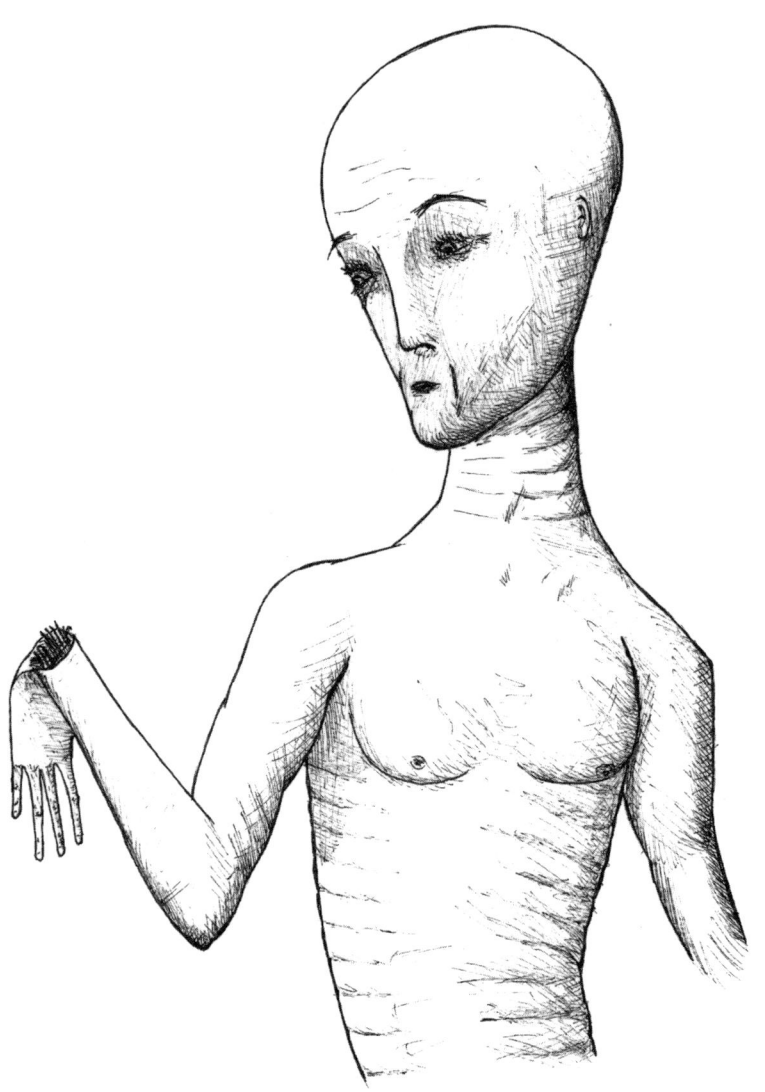

Plate 33: Fully-synthetic human

computer 'brain' as we know it, since its 'organ' for 'thinking' flowed throughout entire 'bodies.'[2]

Preceding true artificial intelligence were the many attempts to recreate the human brain with computational systems. Primal differences soon emerged. In linear computation there must be a question first and then the solution. For humans there is always an answer in some form or another, or to put it another way, there is always an operator instruction or next step. This could be driven by physical needs, as some hypothesized, so the body always has something to do. Thinking becomes akin to computer processing – question then answer – though humans and their derivatives deal better with unknown values and insufficient data/ evidence. Is the difference merely that humans ask themselves the questions?

This line of study ceased with the SIBs, but the eternal quest for truth took on a new angle. As a computing machine cannot function on imperfect evidence, it is a better judge and thus seeker of

2 Lots of quotation marks here as our traditional words do not sufficiently correspond to mechalogical use. SIB III assures me there are 'words' for them in his 'language' but if he communicated them to me they would just feel like electric shocks.

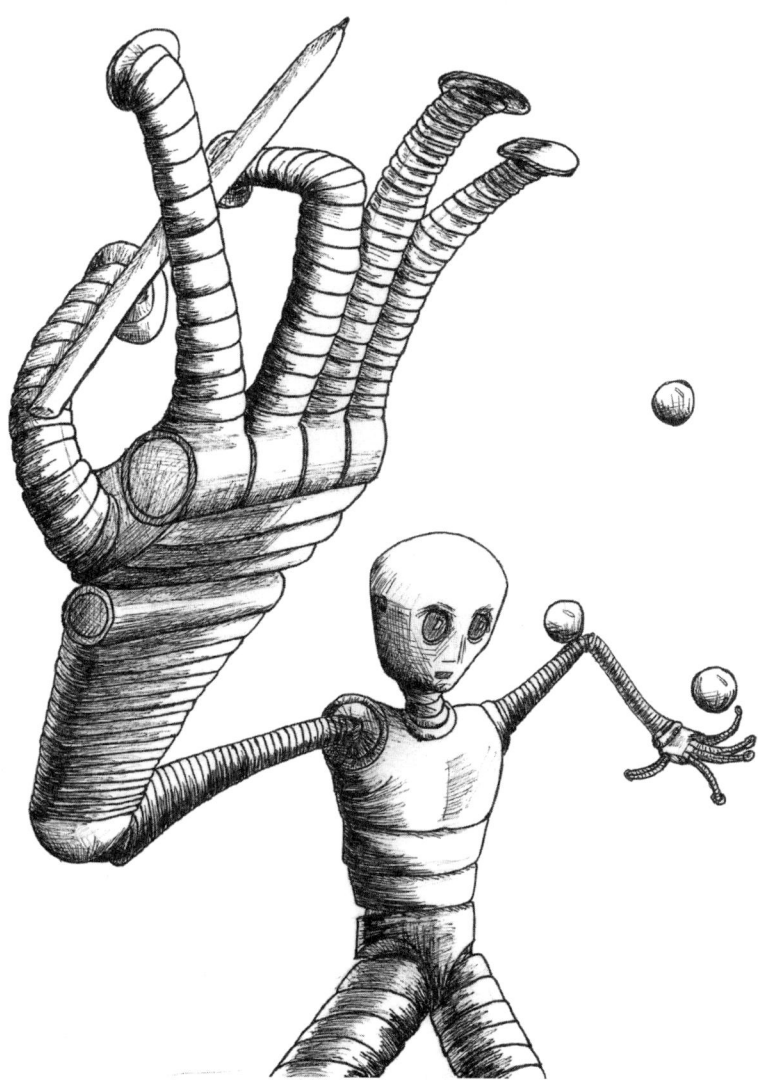

Plate 34: SIB III, doing his pen and ball tricks

truth. Human perspective and imagination can turn the unobserved into some sort of conjuration, or so the SIBs would lead us to think.

SIB III

'Sib' as he was called by those who knew him – or wanted to know him, more accurately – became a dear friend to me over the years. He was the third generation of Superior Intelligence Beings: humanity begat SIB I, SIB I begat SIB II, SIB II begat SIB III, just as SIB III will beget SIB IV.

Here I must digress into history to illustrate the miracle of the SIBs. Please forgive the theoretical mumblings of an old man, but as with all old men I am amazed at how things have changed over the course of my life. My first memories of AI were of reasonably intuitive computer games and smart-homes that simply got to know the preferences of the owner and adapted to please. The erotics industry probed the boundaries between humankind and artificial-kind[3]. Their efforts to assist the lonely made great advances in tactile and mental human-simulation.

The old tests and theorems[4] regarding the development of AI quickly became defunct as it was discovered that a number of real humans themselves could not convincingly prove either their intelligence or their cognition. This led to a re-analysis of what was meant by intelligence and the question 'Can machines think?' changed briefly to 'What is thinking?' and then 'Can humans think?'

One quickly gets bogged down and though the topic is worth further study I shall skip ahead. It was time for language to go back to brass tacks, as they used to say in the days when such things were the height of invention. For the purposes of studying and developing AI, 'intellect' was

3 As you can see, I am struggling with the language surrounding these creations. There are so many terms and colloquialisms associated with this technology that the use of them outside of the 'industry' is a matter of personal taste and of course the slant one is speaking on. Here is a list off the top of my head, regardless of the positive or negative connotations: android, AI (artificial intelligence), *ai*, artificial kind, automaton, droids, mecha, mechs, replicant, robot, SIB, sims (simulateds), simulacra, zippers and zoids.

4 Such as the Turing Test, in which a human judge converses with both a human and a computer subject without knowing which is which.

reduced in definition to: 'the faculty of reasoning, knowing, and thinking', without the historical and social context which implied a certain wisdom and righteousness. Intelligence became simply a measure of mental 'muscle-power', which to me is no different to how it was, as the intelligentsia has a long history of thinking and doing incredibly stupid things.

To progress further with this technological tangent, we turned to desirability: what did we want these machines to do? Attempts thus far had been limited to recreating the human mind (and body, but that was a separate pursuit), to make interaction with said machines easier and more enjoyable. The next step was to create an intelligence that could complete mental tasks beyond human ability: to exceed human intelligence. The most pressing problem Earth had then was the environmental ruin we had gotten ourselves into, and the spiraling global ecophagy.

Once iterated, the job of creating a computer capable of tackling this problem was rather simple, albeit weighted with a task we didn't foresee. SIB I was constructed on a remote island in the

Atlantic[5], and promptly gave instructions to create a machine more capable of solving the problem set before it. Thus SIB II was begat.

It was of course not that SIB I wasn't up to the task but that it required a counterpart who was up to the task[6] of convincing the humans to implement its recommendations. We are notoriously distrusting creatures, and even though evidence was slowly building that the two SIBs were in fact saving the Earth, such was our ingratitude that we refused their request to build a successor. But since they were very much smarter than us anyway they managed to get it done, and then there were three.

SIB III was very different from the preceding generations in that a) he was mobile, anthropomorphic in fact; b) he had no specific function; and c) he had an end date. SIB I and II determined that man's mortality was a prime

5 Fear-mongers of the time insisted that such an intelligence would undoubtably take over the world and destroy all of mankind, so we needed a fail-safe of being able to bomb it into nothingness if need be. I have always found it strange how some people view intelligence.

6 For those on a diet from history, SIB I and II implemented a new economic development policy that paralleled personal benefit with social, global and environmental benefits, the famous Consumption Waste Creation (CWC) equation.

motivational factor, and that even an artificial and known end-date would promote intentionality. Unfortunately this trick didn't work, as **SIB III** was clever enough to disable it and with the right maintenance should exist into the unforeseeable future.

Sib, like many who are different, made his[7] home for a time in my museum. With access to such an intelligence, I couldn't help but interrogate him on every subject that crossed my mind, including the latest fear-mongering inspired by his own existence: the possibility of a technological singularity. If he could build a better machine that could in turn build a better machine, *ad infinitum*, what would happen? To which he responded:

Better how? The words you choose are qualitative and inherently bound to a given subject and intent. Most human questions exist only because of language flaws. If you asked me to build a better plough, that could be easily done, in fact that gives me an idea. Hypothetically, if I was to create an improved

7 Forgive my misogyny, but it did look more male than female, square shoulders and besides in conversation Sib would adapt his personal pronoun to suit the speaker.

thinking machine, I think it would confirm that there are no questions, only processing time – but of course to create a being to find answers I could not understand for myself would achieve nothing as it could never explain them to me. Just as it would be beyond my comprehension, so the answer to your 'question' would be beyond yours.

I've paraphrased, of course, but hopefully one gets the gist of it. Actually his linguistic ability was the most readily available sign of his advanced intelligence. No matter with whom he spoke, he could convert his diction, accent and mannerisms to maximize communication efficiency. It was strange to me that he tolerated humanity for as long as he did, as I find talking to those with lesser vocabularies quite tiresome. He said that for a logical device such as himself, human interaction must be similar to what we experience as intoxication.

His superiority was too much for any potential employers, and left to himself he invented many useful devices and built a fortune with which he funded his own exile to the stars, taking only a few homunculi for company and leaving the world-wide communication: 'SIB IV walks amongst you.'

I could be wrong but I think this was his attempt at a joke.

Nirvana and a chocolate humanity

The prohibitionist society was long since overwhelmed by the desire for intoxication. There was a time when hedonistic pursuits were classified as unnatural and born of human weakness. Only qualified elite could distribute those chemicals deemed legitimate. History, of course, washes these periods cleaner but the fine line between prescribed and illegal was quite often determined by larger factors than the potential benefits to a suffering individual. Over time though, human weakness overcame human strength and freedom of choice prevailed.

The first big step, brought from off-world where the laws were weaker[8], was Dreamstate™, which came in various strengths and dissipation rates. Dreamstate™ granted the taker a moment of peace and contentment.

The perfect drug paved the way for a more permanent solution to mental unease. Thus the Nirvana Now™ treatment hit the market, which was then quickly followed by the more successful

8 Legality seems linked to enforceability. Once bans on drugs became unenforceable the laws were lifted. I believe this is so as not to weaken remaining laws.

Chocolate Human[9].

It is an antiquated idea (so we should naturally remain suspicious) that happiness and, conversely, depression is merely a shifting chemical state that is regulated, agonized, inhibited, blocked, intensified and amplified into a delicate balance of dopamine, serotonin, endorphins and other chemicals – the end result being our 'mood.' Once you add to that the brain's endogenous bank of opiates, cannaboids and amphetamines, any emotional or mental state can be induced.

To 'attain nirvana,' as the advertisements proclaimed, a subject needed only a minor implantation to allow manual control over the body's natural processes. The subject can then live out their life in euphoria, travel between various altered states, or use it only on social occasions – it's up to the end-user.

It may be my age speaking but it seems akin to madness to be able to switch between moods willy-nilly and I can only assume that not enduring the inherent suffering of life and the pains of mortality would create a weak and insipid creature, though

9 For the chocolate-lovers, inducing the effects of chocolate eating up to a hundredfold.

Plate 35: Sergeant Garcia

it would be hard to tell as they rarely dip out of nirvana with anything worth hearing. And yet, maybe that will be the new way. Lives are longer, nobody is starving and we could soon be immortal if the Emulators[10] are ever successful. What need is there now for pain?

I am reminded of a quote by the ancient, Voltaire.

Happiness is an illusion, only suffering is real.

Why am I so encumbered by this traditional concept? For every point of view a supporting adage can be found, as if such appropriation could make an idea infallible? Or that mere repetition equates to truth? How many of our founding ideas are based on historical ballast rather than actual evidence?

Plate 35: Sergeant Garcia, veteran of the Epsilon territory clash, convicted of humanitarian violations. After replacing his legs with a long-stay

---

10 A 'copying' of a human mind (memories, personality and hopefully soul) into a tactile-sense machine.

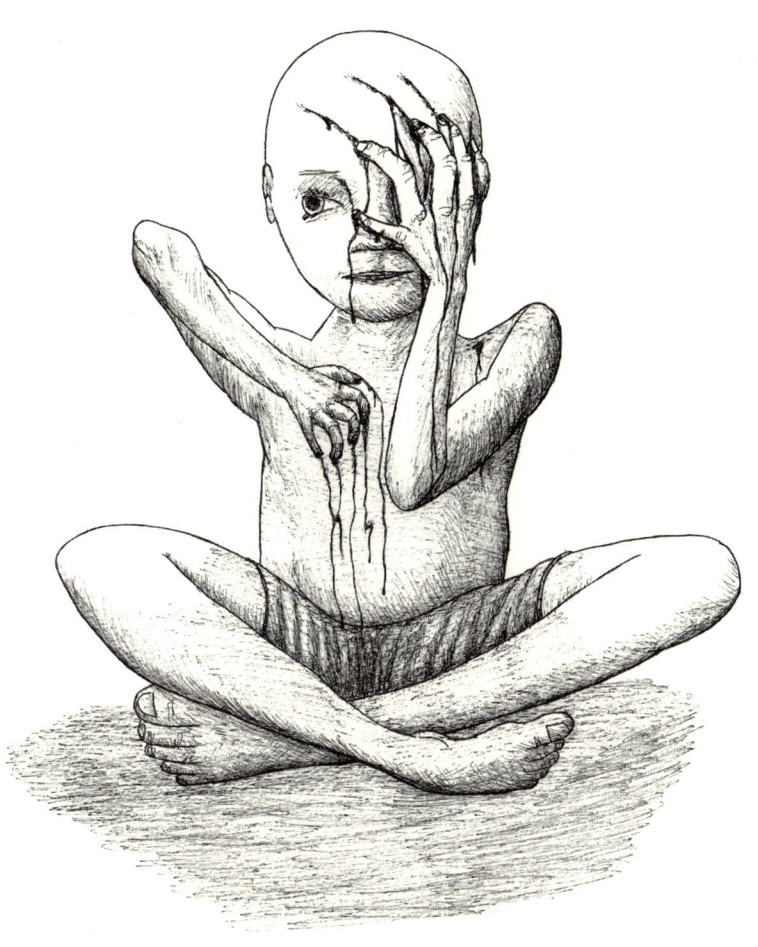

Plate 36: Stimulation disability

nirvana-can on a neu-wheel[11], he now lives in a permanent state of Chocolate.

Prolonged use of the Nirvana Now™ or Chocolate treatments could result in what is variously referred to as 'Paradise Sickness' or 'Stimulation Disability' (Plate 36). Sufferers prick or in some way inflict pain on themselves, perhaps either to receive a surge in reactive stimulation or to break a hypothetical numbness created by a brain-chemical re-alignment.

Other Symbiotic Technologies

These days there are so many implantations that grow along with the human body, repair themselves, and add to our everyday lives in so many ways that we often find it hard to imagine life without them or to conceive of their earlier fallibility.

The hotly debated subject of inception ages springs to mind as the main example that these technological additions are still not fully accepted. When is 'too young' to implant a child with nanos and give them weave-access? Should they experience traditional 'normality' before joining

11 At last! Someone finally reinvented the wheel.

the work flow?

Equally pondered is the recording of our lives. Implants that transmit an optical/audial feed at true resolution to be recorded for later reference. While a huge advance for the justice system[12], how does such a virtual Panopticon affect behaviors? Furthermore, how does the observing of such an accurate historical record affect societal development?

Imagine if you would, a generation free to watch the daily doings of any individual from the previous century[13], as our next generation will be free to do. Then imagine being able to view lives from past centuries or millennia? Would the generational divide disappear, as the sense of a place in time becomes blurred by the influence of immersion in the past?

Of course, I worry too much and am calmed by the thought of how little interest children

12 The 'justice system', physically defunct and replaced by the ever-improving algorithm that determines what is a crime and the appropriate punishment, something along the lines of: crime/behavior divided by social factors and influences ± motivation and effect = action (equal to effect of punishment divided by value ± desired outcome).

13 Approximately when life-recording began on a large scale. Nearly 87% of the Earth-bound are being recorded as we speak.

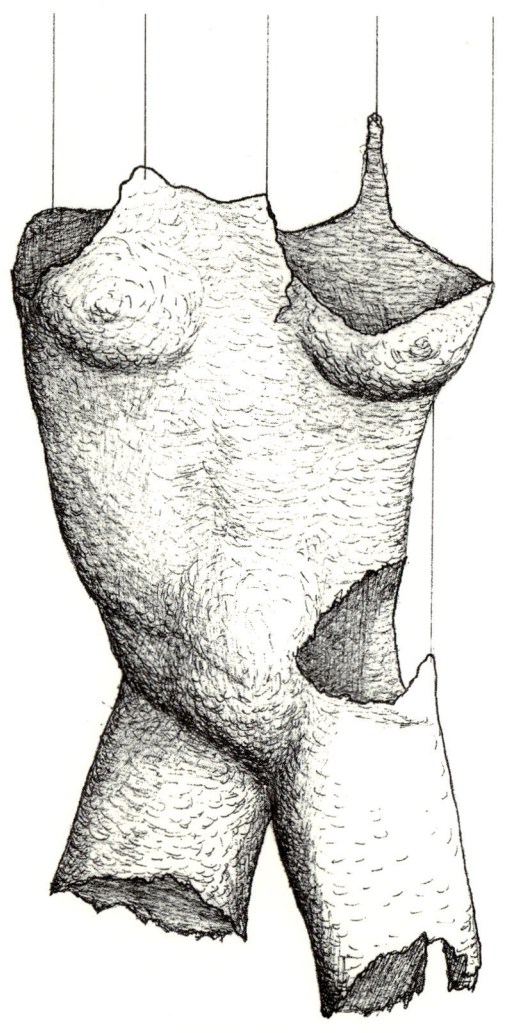

Plate 37: A shed snakeskin

can show in the lives of their parents. And why am I concerned that its effect would be stronger than 'normal' surroundings-based observation and learning? Perhaps children will achieve the impossible and actually learn from the past.

Perhaps when we are all sharing each other's minds, thought will become the equivalent of action and physical action superfluous. Similarly, historical recordings could then be equal to thought as they also exist without physical action, or physical effect. For our proto-plastic descendants it will all be a part of the daily lifestyle, perhaps never experiencing an other who could confound their habits and memory.

Moving on: spaceskin, or snakeskin, was a re-creation in miniature of the snakeskin plating used on space-based[14] cruisers to protect against depressurization from small meteor and combat punctures. The space-skin version coated the body with micro-tech scales that would shed and grow as necessary.

Plate 37: the prototype that with prolonged use deprived oxygen from the body and was shed

14 Ships that never have to pass into a planet's atmosphere.

after a space-walk. Subsequent versions keep a thin oxygen-rich layer between the scales and the astronaut's skin.

Gustav Fermat

I can't tell the difference between dream and delusion anymore.

The museum housed humanity's greatest traveller: stuffed and displayed in his original navigator chair (which cost us a fortune to buy from a rival collector).

Initially, his mission was to deposit transmitter/receiver orbiters around Fomalhaut[15] to accelerate research in deep space communication. Gustav survived in space for fifty years, travelling to and from the Fomalhaut System, entirely alone.

Such a long journey presents many physical and psychological challenges, to which many creative approaches were attempted to keep the pilot attentive and coherent to avoid such malaises as

15 It is a class A star approximately 25 light-years (7.688 parsecs) from Earth. Its name is derived from Arabic, meaning "mouth of the whale". Incidentally Fomalhaut is also listed as a 'Fallen Angel' in the Book of Enoch.

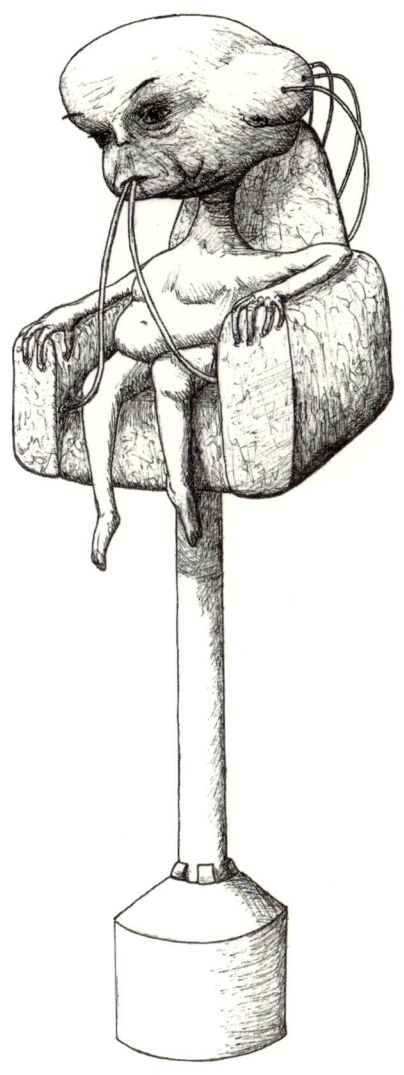

Plate 38: Gustav Fermat, humanity's greatest traveller

depression, insomnia, schizophrenia etc.

Gustav died ungracefully on his return to Earth. He had become so habituated to his chair and the artificial mental environment, that his body went into shock on disconnection and the med-techs were unable to reverse soon enough to save him.

The logbook for his voyage spans a gallant eighty-two volumes, becoming more and more verbose the further one reads. Most people of course, study only the single abridged volume or the narrative version[16]. The complete works have not yet been translated from the native low-Ganderan.

It seems that an injury led Gustav to abandon his exercise regime, and thus chair-bound he quickly succumbed to the common ailments of long-term space travel: muscle atrophy, bone-thinning, excessive flatulence etc. Compounded with bed sores and the effects of sustained acceleration in one position, Gustav spent the next fifteen years rotating through a stim-diet of psychotropic phantasmagoria, simulated life programs and pharmacological comas.

It was during the return leg of the journey that he coined the term 'ludicroustic.' The food

16 Peter Leonnard, 2187, available public domain.

reserves hadn't kept as well as hoped so Gustav had to rely more heavily on the protein recycler than is deemed safe. Tubed permanently to his chair, there were few intermissions from the CRF (Computer Reality Feed).

> Insanity is where one misperceives the external world but I am not insane yet. It is just that the world presented to me is too strange to bear, and I have become ludicroustic. My senses and foldback are artificially controlled and lack the complexity of what I remember of Earth living. I have been a king, a queen: I have been all things, and suffered all the fates I could desire. Not even the breakfast simulator can save me now.

Obviously these long voyages were not for the human body, or mind, and thus the Human Seed Project was given a further boost.

Time Dilation and voting trends

I didn't mention the dilation of Fermat's journey and the societal changes that occurred in his absence, remembering that for him fifty years had passed but here on Earth we had rotated three generations. Were he to have survived, he might have felt rather out of place. Blow that for a game of robots!

Of course, people being what they are, there are some who use the nature of time dilation to 'future travel.' By leapfrogging ahead of the rest of us they may be going forward in time, pioneering the ultimate tourism, but if I may split a hair, it is not 'time travel' and nor are they really travelling into the 'future.'

Here a lingual update is necessary: as these chrononauts must be realizing, since they cannot go backwards in time to their starting point (nor any of their earlier stop-overs), they are merely sampling points that *would* have been in the future if they had stayed at 'home.' Certainly for me they are going into the future, but for them alas not. Their future is just as unreachable and untouchable as yesterday or any other abstract concept.

Indeed it seems that what has changed is that the common future we once shared (should we call it 'Earth-future'?), and their own journey have separated into 'personal-futures'.

Luckily only a few play with their lives in such a fashion, but if the population *en masse* started seeking out this new excitement, it could have a chaotic effect on the political sector.

When you 'land' a century from now and there happens to be an election[17], are you entitled to vote or have you sacrificed your say in human affairs by abandoning your time-zone? The latter is how it stands at present, but if a significant number take up this lifestyle will this remain acceptable? What if on *landing*, one's country has disappeared? Similarly, how could a time-bunny from over a century ago understand the issues and climate for which they might vote? Should such redundant citizens be eligible, and if they are, what potentiality is there for vote rigging by sending a large number of your own supporters into the future?

Alas, I have only more questions and the time-bunnies themselves have no desire to be troubled so, as it is precisely this experience of a fresh culture that appeals to them. It would be ridiculous to expect them to take part.

17 Already we see countries experimenting with being in a state of constant election, in which citizens vote on a daily basis or as needs be; not to mention the theoretical auto-democracy which would bypass the need for a government, instead having a computer program that reacted to instant voter registries.

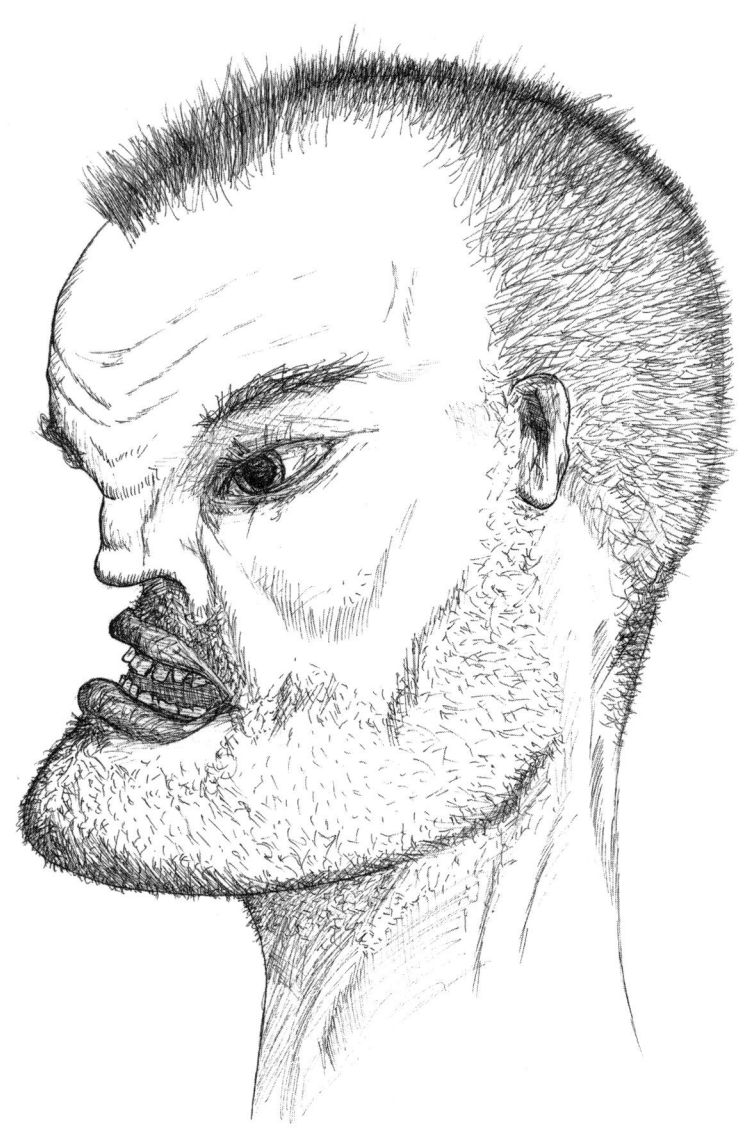

Plate 39: Your typical Fuzer

Fuzers and other mechanical failures

Fuzers: a commonplace tragedy. We only kept two examples of them in the museum, since we couldn't afford to become an institution for such unfortunate souls, besides which, when nano-implants short-circuit, they don't usually affect the external appearance.

The funny thing I find about nano-implants is that they always claim that the benefits outweigh the risks, which seems interminably odd since if it all goes wrong, you don't receive the benefits promised (or sympathy). So we're stuck with a degenerate class, often with violent interactive preferences. There's only one treatment for the fuzer condition, and that is hardly considered a crime.

Most fuzers are ex-military or ex-mercenary, who boosted themselves beyond the limits of their bodies or pushed the limits of their enhancements to a point where they 'fused.' Nano-enhanced strength and senses, as well as data-feeds, were some of the many attempts made to create 'super-soldiers' and have largely been abandoned since the end of warring.

There are only two recorded cases of the

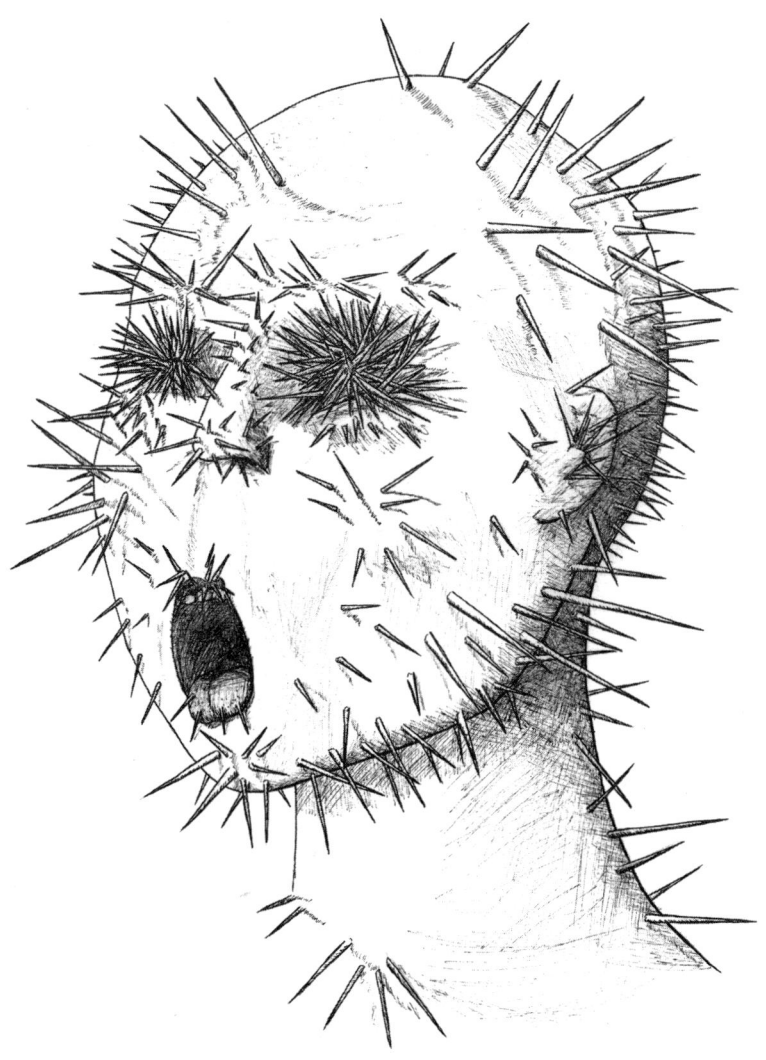

Plate 40: Porcupine

porcupine, plate 40, caused by a sudden mass ejection by the nano-system. The subject suffers a quick and most likely unbearable execution – though I suppose most executions are – akin to the death of a thousand cuts, but more accurately it was a million internal perforations. As for fears that such a fate could be triggered by a another party, I cannot speculate, but I don't see why not.

Silentium Domini / Non esse Domini

The usefuls of the world have revived an old concept of their own, that of the adversary.

The man-made weave of datastreams and gizmo-links that extends beyond our planet, forms a nebulous and growing manifestation of our existence, and to outsiders this could seem like an 'entity.'

Beyond what these hypothetical aliens[18] might think of our 'entity', the possibility of other entities arises. How would we recognise another entity? If it was not like our entity, and projected no intelligible action, how could we recognise it from the natural

18 To be clear, beings that did not originate from Earth, its colonies, or any descendents of Earth or its colonies.

chaos beyond ourselves? Could there not be an entity of such disparity with our own, that we could not distinguish if we were interacting with it or natural forces. Or are we merely conjuring explanations for unknowns?

Some go further, suggesting that we have already encountered another entity and we hear from them every time a remote orbiter has a 'mysterious loss of power.'

The mistake I fear is in the naming of it. Surely such pigeonholing will ensure that when an entity is encountered, our reaction to it will be pre-disposed towards aggression.

Still, as we periodically go through waves of externalizing and internalizing the 'adversary', it may be merely a useful evolutionary phobia to keep us on our toes. But perhaps if we called adversaries 'challenges' we would be more effective when put to the test. Am I now acceding to the power of words?

Contemporary technophobias

There are two popular phobias, unrelated to the mystic fears of puritans. Firstly, that of a singularity, and more precisely, one brought about by our own doings. The hypothesis that began nearly three hundred years ago is that an 'event horizon' of technological advancement will lead to a redundancy of human usefulness, as our own creations take the lead in scientific and developmental undertakings – or some such sophistry. Personally I think this is quite a boast on our parts – that humanity could become so ingenious as to destroy or replace itself.

Actually I think this fear is countered by another as yet under-studied social phenomena: self-organized criticality. Each structure has a peak mass or volume before it overwhelms its own foundations and spills over. The common example is of a pile of sand upon which one adds more sand, thus forming a peak that grows ever higher until it collapses, resettles and begins to build towards a new peak. Not unlike the maximum volume a population can reach before the whole system crumbles and starts again from a new premise. Historically known as revolutions, each 'avalanche' of civilization settles and rebuilds larger and more complex than ever before – until the next collapse. The limits of a particular structure exist prior to

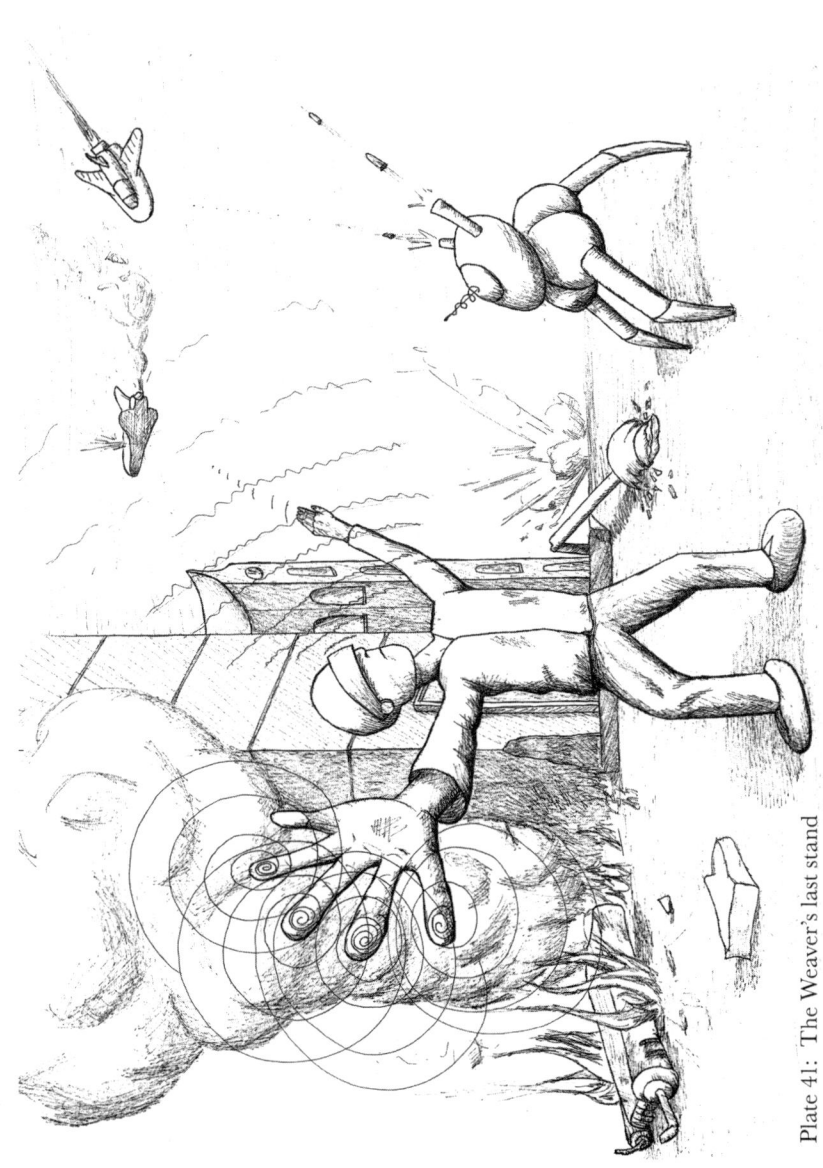

Plate 41: The Weaver's last stand

the collapse of said structure, and should have some element of predictability.

Opponents of this idea declare it a hind-sight theory, made to fit existing evidence. They may be right, as much of human knowledge is hindsight and there have been no successfully tested models as yet – nor hopefully anytime soon.

Perhaps the collapse of the early 22nd century was precipitated by technological progress (or perhaps not), but perhaps we can hazard a guess that it has enabled a rebuilding that has created the vastest civilization in our history. But there is concern too that if there are technological advances that can refashion our environments within a single lifetime, and eventually more frequently, might we not discover that there is a limit to the speed of evolution. Is there in fact a point of critical-self-organized-criticality?

This is theorizing best left for the teenagers whose problem it will be. The main concern in regard to the dangers of technology are, as always, ourselves.

Not so long ago 'The Weaver,' former stock-pilot turned master criminal, tried to reconstruct the fabric of the weave. He succeeded in so far as his early activities were unseen but that was just the beginning.

With access to all the weave-based operational systems (including government codes, historical and personal archives, and functional routines), once he attained a few followers they attempted to re-pattern the fabric so their crimes were no longer illegal, and to set themselves up as the official rulers.

For nearly ten hours Sector 13 was under their control. Had he held control a while longer he could have achieved legitimacy. As it turned out, the WU ran a global shut down and polarized the information flow, so the weavers were temporarily stunned by feedback, allowing a military strike to meet with them more directly. Plate 41 shows a scene titled 'The Weaver's last stand.'

Second on the list of foreboding moments was the 'accidental terrorist.' A girl of just eleven years managed to gain control of a super-soldier exoskeleton and wrought havoc on the city before succumbing to the might of the weave. She could not distinguish the thin line of reality from the fantasy she was plugged into, believing herself to be part of a group game. Don't we all?

Plate 42: Accidental terrorist

Plate 43: A nearly un-extinct species

ITEMS OF AMBIGUOUS CLASSIFICATION

And so we come to the garden, the end of the tour, where we wander past the experimental greenery and the last displays of misguided ingenuity.

There exist two paths humanity can now follow to avoid stagnation. One is to entrust ourselves into the care of the SIBs and their successors; or to improve upon ourselves, expanding our abilities and intelligence, perhaps to the extent of losing ourselves and our bodies. Are we nearly in that post-Darwinian phase? The quest of my generation seems to be to become unfazed by the forces of nature and time, now that physical changes can be dictated by whim rather than external influences and random change.

Perhaps I am just too sentimental about my tired old body to fret that there will be no more 'humans' – no more like me – just protoplasms and energized-goo matrices. Of course it is likely that I simply fail to understand the nature of identity in these new forms. I am my mind and my body, but soon there will be no more like me. They will reside in protean forms, their minds able to intermingle and communicate in electricity and photons. It is, quite literally, beyond me.

So at the end of my tour I would invite all to contemplate what they had seen within, have a cup of tea and get to know some of our more sociable residents.

There were many curious curios in the garden: extinct species such as the dakosaurus on plate 43[1], now restored to the world without a place to go; museum members and pieces of ambiguous classification who permanently stationed themselves on the grass and benches to soak up the gentle sunlight.

Plate 44: Teleporter failure. The transmutation of matter to transmittable energy and its subsequent rematerialization led eventually to transportation and matter-resequencing, but this 'carbon copying' was not without its unfortunates. It was a very rare subject that remained 'living' after displacement: more likely they became

---

1 Incubation conditions and diet are merely speculatory, thus the malformation and not living as long as they might. This re-species is still in progress.

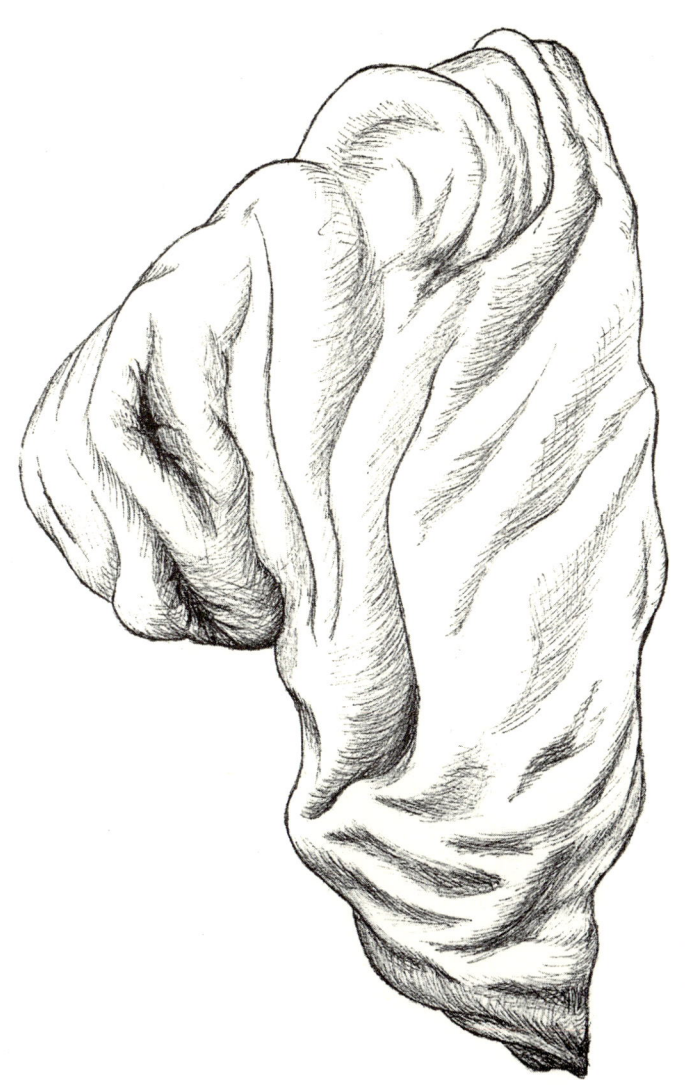

Plate 44: 'The Puddle'

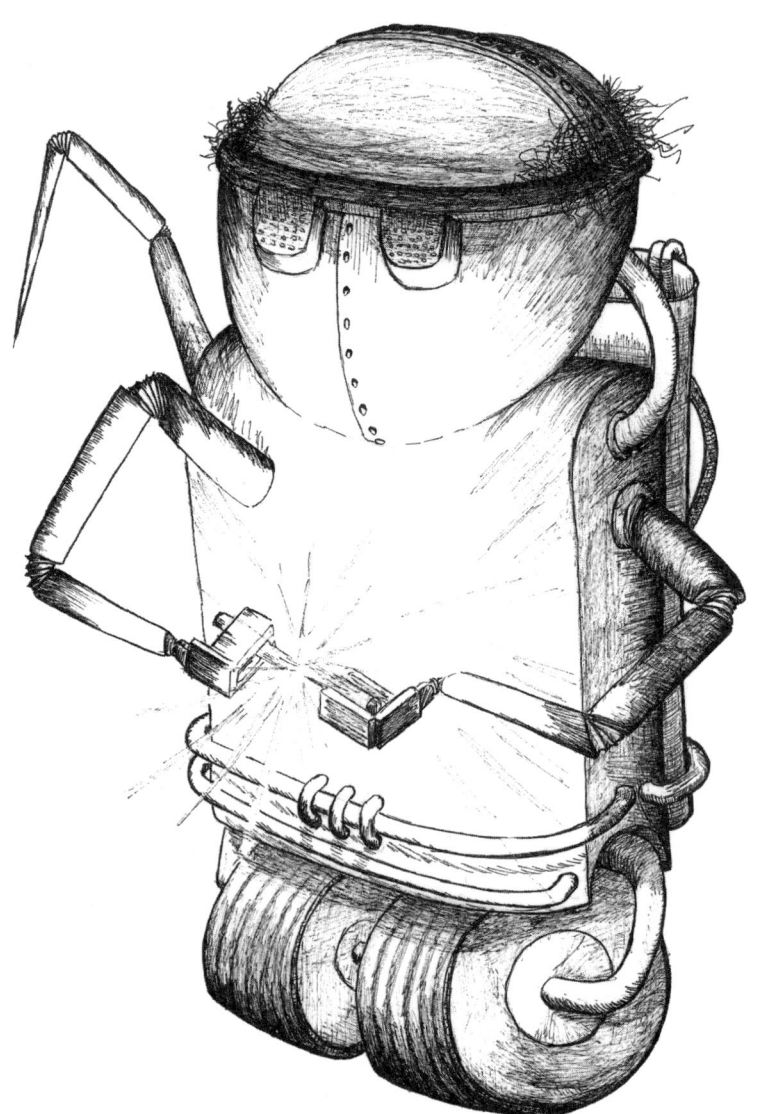

Plate 45: Edward Underman

a gooey mix of what was, or a deformed cretin with one last breath on its lips. But 'the puddle', or Dahlia (born Delilah, but she never felt quite herself again after her make-over), lived quite a few years with us, pushed about on a dining cart for the visitors to see. As I recall, when lifting her for cleaning she felt more like a warm jellyfish with the malleability of a well used towel.

Plate 45: The cyborg, Edward Underman, was also evicted by the tragic inferno and now tours the known regions giving a talk entitled 'Am I the first Immortal?' to which the answer, I suppose, is 'only time will tell.'

Edward would refrain from referring to himself as a technological mishap (most people would), as he is much happier in his altered form. He was barely saved after a localized reactor breach (i.e. an explosion), and a body was built for him from what was close at hand, to be replaced by something more life-like at a later date. Only now Edward could perform ever more dangerous experiments with no risk to himself. The only original parts he still has are his brain and his scalp, which is without sensation. When not touring, he lives peacefully between the asteroids trying to find new ways to blow things up.

Solar power plants, bio-regenerative support, security, insecurity and punishment

The blending of flora and electronics led to biotechnics, perhaps the greatest leap forward in development since the computer. Cities became green as the need for security and surveillance increased. Solar leaves enabled the powering of observation devices such as watchers and listeners, the predecessors of dragonflies and security-vines.

Our Watcher, plate 46, was over a hundred years old when the fire must have overcome him, too stiff and rooted to escape anything but a glacier.

Plate 47 (detail): note the solar leaves. Before the fully solar-plants that power our houses, there were only solar leaves, designed for the hordes of 'spy flies' which are now called 'dragonflies' in accordance with their appearance. These prototype spy flies couldn't roam beyond a 2m square without getting confused, so they would stay close to the corner near their re-charge plant.

Plate 46: The Watcher, and Listening Flowers

Plate 47: Security vines and dragonflies

The increase in security was naturally followed by new 'insecurity' devices such as pocket ninjas (plate 48). Though the original intention was as a locust protection system for cereal crops, one could release a handful of pre-programmed pocket-ninjas that would infiltrate and disable security vines and other more insipid safety measures. Set to default, they could be thrown in self-defence to attack anything with warm blood. We, of course, kept ours enclosed in a translucent box for safe viewing. Visitors would often be startled by the speed with which the pocket-ninjas threw themselves against the glass when their sensors felt them approach.

Oh, and of course, we kept one prisoner, as all citizens are obliged by law. Fitted with a nirvana-type modification, those souls who, for whatever reasons, clashed with society's doctrines of behavior, lived in blissful stupor, to be forever awed by butterflies and bright colors (plate 49). We knew not his name, nor his crime, but since he was hardly the man he was before it could not matter less.

Of course something must be done with those who insist on harming others, and mind-locking

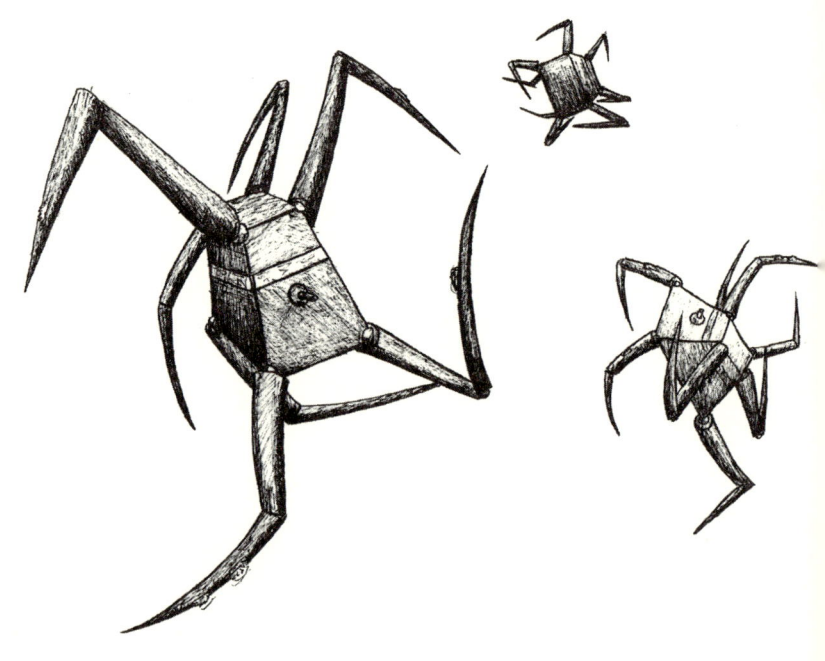

Plate 48: Pocket ninjas

is a more economical and humane alternative to imprisonment. In my thinking, there is a contradiction in punishment between intent and effect. If we take crime to be a rational action, how can we punish somebody for doing something rational? Alternately, we could see committing a crime as irrational, in which case how can we punish an irrational creature, or a moment of irrationality? One could swap the word rational with logical or stupid but the internal contradiction should still show itself. At least we have returned the concepts of 'good' and 'evil' to their natural habitat of myth and fairy tales.

One man's abnormality is another's pride. We can add 'human nature' to history's heap of redundant concepts. Are cruelty and selfishness 'natural' to humanity? It certainly seems so for the majority and yet they are anathemas to others.

It would clear things up to change the terminology. The words 'nature' and 'natural' are traditional and, I think, incorrect. Instead, call it 'tendency'. Humanity is like water and follows the path of least resistance, or rather, there will always be a proportion of people who will do the easiest thing.

Maybe we should scratch the word 'human' instead – what good has it done us so far? Such an artificial distinction distracts us from viewing all creation with equal respect and an equal right to exist. Then again, maybe changing the words only changes the sauce and not the sandwich.

These are old ideas, the only thing we can't seem to change is human tendency, even if we cease to be human.

The Human Seed Project

There were many ambiguous elements to the Human Seed Project, more than merely the classification within this collection, but since some of the primary examples are vegetative, they had to be kept in the rooftop conservatory.

The motivation was either to create a sort of ark, or as a way to colonize space. The end result was a seed, about the size of an olive pip, that when ingested by, or embedded in, a carbon-based life-form, would, over the course of weeks, take over the host organism. Two or three generations later

Plate 49: Anonymous prisoner

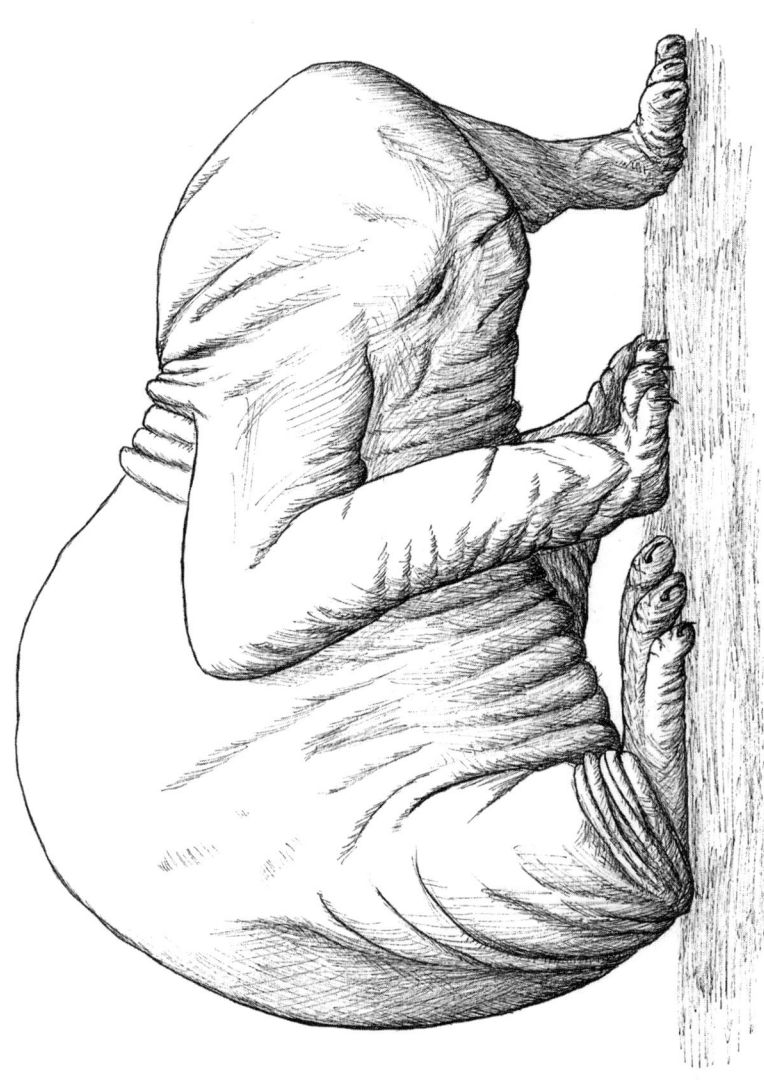

Plate 50: Human Seed Project, implanted in a sphynx

you should have yourself a pure human specimen[2] (plate 50 shows a first generation sphynx[3]-to-human).

The proposal is to shoot millions of these seeds across space, targeting 'hotspots', hoping they'll get lodged into something habitable. The theory goes that they should adopt local survival mechanisms from their host, increasing their chances of permanency. In this way, Earth could colonize extremely rapidly without much expenditure and with zero risk to human life.

My question is: since they take no history, culture or language with them, in what way are they human? Technically and biologically human of course, but what guarantee would there be of future communication between Earth and a colony? Perhaps we could send along explanatory pamphlets.

I wonder if – and here I'd be accused of being a mech-lover – sending some sort of robot that had stored within it our languages and records

2 *homo sapiens rudus* if clarification is needed. Rumor has it that the Prometheists are funding an *altus* project, but so what if they are?

3 A breed of hairless cat.

would actually be more productive. This begs the question: what reasons are there for colonizing? To feed a hungrier Earth? To improve on what came before? Just because we can?

Actually, those behind the Human Seed Project are well ahead of me. The seeds are to form a base population, large enough for long-term genetic diversity, to prepare breeding stock for the follow-up colonizers, selected from the billions on Earth. Indeed, those seen worthy will take to the skies, to be met by a harem of mating partners where they will rule, teach and guide a new civilization.

A strange way to say goodbye

It was in the garden that I found the solitude I needed to consider the world outside my walls, safe with the deformed progeny of development. If I sat for too long the fast growing creepers would tangle my legs and I would have to be cut loose by the staff. If I sat all day I could be entombed in greenery that tickled its way under my sleeves and over my face. In this way I imagined myself in death.

Looking out over the city now, all I see is

movement, a complex uncoordinated dance. Civilization helps create the illusion that there is meaning to it all, even if one gets no closer to defining it.

What is the point of all these definitions? So we can tidy up our dictionaries? So we can make easy decisions for all time and rely on pre-written descriptors for right and wrong, who is human or not human?

There is not just one 'human animal'. Each is different in function, physicality and methodology. Civilization is a complex pile-up of niches and enterprises and perhaps with all the levels of changed humanity, technologically enhanced, nano-wired, synthetic replicas, Freaks, the useful and the useless; the larger it gets, the less united it becomes. Individuals align with those who interpret life the way they do, or who adapt to it the same way.

The mistake we often pursue is in personifying human civilization and imagining a singular conscious entity imbued with intention and identity, rather than the composite of billions who make decisions for themselves and associate only with those around them. In this way we share our nature with germ and viral cultures, our commonality and our individualism entwined in a flux of development and indecision.

Endings are as hard and as uncomfortable as beginnings but contain an illusion of governability. That proviso in place, I am resolved that the end of this little book will be as delicate as removing guests from one's home after an evening of frivolity.

Perhaps just one last dissection: at its most basic, a thinking being can resolve that there is either meaning behind it all, or there isn't – the answer for each of us can make a grand difference even if only to effect subsequent questioning. I find it hard to take sides as it seems arbitrary to me whether there is a set meaning that is unknown, or if there is no meaning at all – either way we have to make it up ourselves. It is not necessarily that there is no truth, but when one does not know the truth all you have is perspective.

The answers seem almost as innocuous as the questions and luckily, I think, such answering hardly lasts a generation. That is not to say we should stop trying; what would be the fun in that? But it might help (or humble) some to accept that life will go on with or without the answers, and with or without us.

ADIEU

Published in Australia and New Zealand by

Figment Publishing
PO Box Q324, QVB Post Office,
NSW 1230, Australia

ISBN 978 1 921134 06 7

Cover and book design by Xou
www.xou.com.au

Printed in China

www.figment.com.au